Burning Down UNESCO

A Guide to Innovative Fundraising

Howard Burton

First published in 2012 by Open Agenda Publishing Inc.
Second edition, with epilogue added, published in 2021
Copyright © 2012, 2021 Howard Burton

Open Agenda Publishing Inc.
7 Mead Court
Toronto, Ontario
M2L 2A5
Canada

All rights reserved.

ISBN: 978-1-77170-135-8 (pb)
ISBN: 978-1-77170-131-0 (ebook)

ABOUT THE AUTHOR

Howard Burton is the author of three other books: *Exceptionally Upsetting: How Americans are increasingly confusing knowledge with opinion & what can be done about it*, *First Principles: Building Perimeter Institute* and *Letters From Languedoc*.

He holds a PhD in physics, an MA in philosophy, and was the founding executive director of Perimeter Institute for Theoretical Physics from 1999-2007.

In 2012 he founded Ideas Roadshow, which was broadened into Ideas On Film in 2021. He has created dozens of films and edited over 100 books based upon detailed, long-format conversations with internationally renowned experts in a wide variety of different subject areas.

Visit www.ideas-on-film.com for more details on all books and films. He lives in France.

Contents

The Intrepid Idealist 8
Penetration 17
The Education Begins 26
2+3=8 and Other Revelations 43
A Strategic Error 59
Burn Before Writing 76
Historical Investigations 80
Drowning in Humanism 96
Lighting the Match 111
8.6% of the Way 131
The Problem 134
Concrete Utopianism 101 141
Harnessing A Crisis 151
Hope Springs Eternal 163

BURNING DOWN UNESCO

A GUIDE TO INNOVATIVE FUNDRAISING

CHAPTER 1

The Intrepid Idealist

There's a feeling of unease right from the very beginning. Having successfully penetrated the clunky, '70s-style security system (a sentinel posted outdoors to redundantly point me towards the reception desk inside the front doors), I surrender my passport in exchange for the right to pass through the turnstiles and stand in front of a dreary bank of elevators adorned with interlocking white, amorphous shapes positioned over a series of raised letters that proudly declaim: *Cultivons la Paix*.

I reflexively narrow my eyes sceptically at this, my first direct encounter with what I later associate with a typically UNESCO phenomenon: strident declarations of the obvious tinged with an unmistakable whiff of sanctimony and self-righteousness: *Congratulations: you are now waiting for your elevator in an officially anti-war building, safe in the knowledge that you are comfortably surrounded by fellow morally elite, peace-loving souls who unhesitatingly support the frustratingly controversial goals of eradicating world hunger, educating children and opposing genocide.*

Somewhere else, one imagines darkly, there are mirror institutions which serve the interests of the unbridled famine-fixated war-mongers (The Pentagon? The International Monetary Fund? The OECD?)—places, presumably, where banks of elevators are ominously equipped with black sculptures that are defiantly emblazoned with messages like: *Cultivons la Guerre.* Just not, of course, in French.

But I am getting ahead of myself.

It is late 2010 and I have just been hired by UNESCO's Bureau of Strategic Planning to begin a five month consultancy. Officially, my charge is "to explore and potentially develop a spectrum of 'innovative financing' mechanisms to assist UNESCO in its quest to adapt to a changing economic climate", as the euphemism so tersely goes. Supported by annual contributions from 193 Member States, many of whom are showing mounting reluctance to make even inflation-adjusted increases to their annual levies, UNESCO now finds itself under considerable budgetary strain and there is a real and pressing financial pressure to go beyond the current paradigm by seeking out new and different (i.e. "innovative") funding mechanisms.

As someone who had spearheaded the creation and development of a new scientific research institute that successfully managed to unlock close to $350 million in both private and public funding, I might well be just what the doctor ordered to enable the United Nations Education, Science and Cultural Organization to climb out of its current financial morass. Or so, presumably, went the thinking by the powers that be in Strategic Planning-land who had decided to engage me.

While it was hardly my childhood dream to become a professional fundraiser, it's worth stressing at the outset that, placed in its proper context, I have a good deal more appreciation for the whole process than most people I know. Don't get me wrong: the tawdry mechanics of kowtowing before megalomaniacal philanthropists, arm-twisting stodgy bureaucrats, or desperately searching for common ground with rampaging corporate interests holds about as much appeal to me as a double root canal, but it's simply part of the price to be paid to ensure the creation or continuation of essential programs.

More generally, I've come to appreciate in my more philosophical moments that, like early morning workouts in unheated swimming pools, the inherent unattractiveness of the fundraising experience in no way diminishes the fact that it is, deep down, good for you. Very good, in fact. After all, there's nothing like the flinty reality of budgetary allocations to focus the mind on what has succeeded and what hasn't, what is out-performing and what is under-performing. And there's nothing like having to make the case for additional financial support before a potentially unsympathetic audience to impose a more independent perspective on one's strengths and weaknesses.

In short, a candid examination of funding priorities is inextricably tied to the entire concept of strategic development by forcing an objective assessment of the relative merits of an organization that might otherwise be postponed, perhaps indefinitely.

The private sector understands this instinctively as the unrelenting economic Darwinism of the marketplace blithely imposes its presence on all participants. Rigorously objective self-assessment is a must in a world where current success is no guaranteed predictor of future growth, and it is well understood that the competition will unceasingly toil away at targeting your weaknesses to gain more market share for themselves. Today's complacent are tomorrow's dinosaurs, and any period of falling profits or dipping stock prices will swiftly result in the baying of investors for a comprehensive strategic re-evaluation and/or the heads of those in charge. Adapt or die. Or sometimes, sadly, adapt and die. One never knows.

But however challenging it might be to discover the winning strategy, at least in the corporate world the litmus test for success is brutally clear: make money. There is some natural ambiguity between the short term and the long term (how much should one invest in research today in the hopes of generating greater revenues tomorrow?), but on the whole, it is not terribly difficult to discern which companies are currently succeeding and which ones are struggling.

In the non-profit universe, on the other hand, the definitions of success and failure are invariably much more subtle, and often considerably more opaque. Multinational educational and cultural organizations are not in the business of generating profits or increasing shareholder value and thus must work considerably harder at developing clear and objective assessment mechanisms. How to determine the relative impact of a full spectrum of educational programs? How to measure an increase in scientific capacity amongst under-developed states? Sometimes the programs that are the most expensive to maintain actually have the greatest overall impact. Sometimes they don't. It's hardly a straightforward business. Yet it is indispensable. A company that has a poor quarter might be required to lay off workers. A country that can't properly educate its citizenry condemns itself to perpetual poverty.

And there is another important structural difference between the two worlds. In the corporate sector, a company that consistently underperforms will ultimately die. If one can't meet one's sales quotas, if one's market share begins to shrink below sustainable levels, then the enterprise will be forced to make significant readjustments. And if this restructuring doesn't right the boat and no last-ditch solutions are found, it will eventually go under.

Multinational organizations, as everyone knows, rarely go out of business. While this is in some ways a good thing—providing the necessary continuity in programs, policies and infrastructure for those involved—the absence of this formidable evolutionary threat inevitably generates a sense of smug complacency within the organization, twinned with an increasingly inefficient bureaucracy as any new "reform" that somehow manages to see the light of day invariably finds itself doing little more than structurally re-iterating the status quo.

To this rather dispiriting picture, I'm convinced, fundraising can ride, rather surprisingly, to the rescue. If budgets are withering across the board, then the prospect of securing new, additional resources enables a bracing sense of competition and an opportunity to candidly assess past performance. Rather than simply dawdling along according to the long-established status quo of algorithmic disbursements, one is now forced to say explicitly: *We need extra resources to do **this*** and *Without more money, we won't be able to do **that***, while making a clear, penetrating case for the objective merits of specific initiatives and candidly assessing what has gone wrong in the past. Put another way, the process of engaging in strategic fundraising initiatives forces an organization to step back and frankly assess the question: *What do we do that is actually worth supporting and how do we find a way to do much more of that?* And so it is that financial constraints ironically enable substantive progress, forcing honest self-appraisal and turning otherwise detached senior leaders into active champions, unanimously converging around established meritorious initiatives.

That, at least, is the theory.

But of course, for most organizations none of this happens at all. Instead, there is a great deal of knee-jerk caterwauling about how they are being criminally underfunded by an evil combination of heartless, unthinking bureaucrats and populist politicians, complete with misty-eyed reminisces of the glory days of yesteryear when sufficient funds were unthinkingly transferred into their capable hands without any of this demeaning business of being forced to justify their allotments in terms of efficiency or efficacy or any of those other humiliating buzzwords.

For these people, fundraising is naturally viewed as nothing more than yet another depressing consequence of living in our lamentably unenlightened age: grovelling before the modern-day Medicis (philistine billionaires and depressingly well-funded private foundations) in an effort to extricate relatively paltry sums from those who unjustly find themselves with far more money than they have a right to, and a good deal more than they know what to do with.

I've seen this sort of reaction many times before when I was setting up a theoretical physics institute. You might naively think that the advent of hundreds of millions of dollars of new monies towards basic research and scientific outreach would be met with considerable enthusiasm by those who publicly aver the importance of scholarship and intellectual inquiry. You might equally imagine that other more entrepreneurial types would seize on this precedent and actively construct additional innovative partnerships between the private and public sector in a wide variety of exciting and impactful areas. You would be wrong on both counts. As a general rule, it seems far easier to shake one's head and envy those lucky bastards with their sugar daddy than to reflect on what lessons are to be gleaned about one's own organization that was so inexplicably overlooked.

But back to UNESCO: budgets were tight and fundraising was desperately needed. Good. I was engaged as a consultant with the Bureau of Strategic Planning and thus might well have a sufficiently broad-based remit of investigating possible ways forward. Very good. Given my views on the strategic importance of fundraising, this seemed completely appropriate and clear evidence of a profound resonance between my values and those of UNESCO. After all, where better to make strategic impact than the Bureau of Strategic Planning? I was clearly in the right place.

I was trying hard to be positive. The truth was that my expectations weren't terribly high. Before taking the consultancy gig I contacted several people I knew who were well-versed in science and education policy in a multinational context, and the preliminary feedback wasn't entirely encouraging.

"*UNESCO? Why on earth would you want to work **there**?*" was, it must be admitted, a fairly common reaction.

A shrewder soul than I might have taken this as a dire warning of things to come, a shot across the bow of my optimistic little boat, a recognition of an intractable dysfunctionality well beyond the scope of whatever salutary bit of objective reckoning I might be able to bring to bear on the situation.

But I am a particularly stubborn fellow.

True, on the very few occasions when it had forced itself on my consciousness, UNESCO had long struck me as patently irrelevant. The fact that, during the better part of a decade spent building a new scientific institution with a strong educational and cultural mandate, I had precious little reason to ever constructively interact with the United Nations Education, Science and Cultural Organization was in itself, it must be admitted, a rather ominous sign.

So, too, was the fact that the place had only very narrowly avoided selecting Farouk Hosni as its new Director-General in September 2009, finally plunking (31–27) in favour of Irina Bokova, the former Bulgarian Ambassador to France. Mr Hosni, you might recall, was a former Egyptian Minister of Culture, and strongly endorsed by no less a credible figure than Egypt's former strongman Hosni Mubarak. He had garnered not inconsiderable media attention by repeatedly voicing stridently anti-semitic remarks, from boasting how he had assisted the Achille Lauro hijackers escape the Italian authorities, to pledging a ritual burning of any Israeli books that had somehow found their way into Alexandria's new library. After her narrow victory, Ms Bokova took pains to publicly express her sincere friendship with Mr Hosni who, for his part, blamed his defeat on the "USA and the Jewish Lobby". All and all a rather unorthodox way of cultivating the peace, one might say.

And then there was that whole sordid business concerning UNESCO's Obiang Nguema Mbasogo International Prize for Research in the Life Sciences. Mr Obiang, the dictator of Equatorial Guinea since 1979, is Africa's longest reigning leader and widely regarded as one of the most corrupt rulers on that continent, which is no small achievement as these things go. Back in 2008, Ms Bokova's predecessor, Koichira Matsuura, had announced Mr Obiang's generous $3 million endowment fund to establish the prize, proudly sending out letters to government officials around the world to solicit suitable nominees for the award. This provoked, rather unsurprisingly to anyone outside of the corridors of UNESCO, a storm of protest from the international community followed by pathetic vacillating by UNESCO officials, as they visibly blanched at the prospect of returning any monies in their coffers, however ill-begotten. When Ms Bokova once again announced that UNESCO was moving forward with the process of granting the award in April 2010, the outcry was predictably loud: US Senator Patrick Leahy wrote to Ms Bokova and pointed out that the $3 million the president was using to endow the prize in his name was most likely stolen from the citizens of Equatorial Guinea, while *The Economist* bitingly suggested that other UN agencies might want to follow suit by offering similar "innovatively financed" awards like a Robert Mugabe Prize for Agricultural Productivity and a Sergio Berlusconi Prize for Sex Education. Faced with mounting global ridicule, UNESCO took a customarily strong stand and once again resolved to defer making a decision on the matter for another couple of years.

In short, there was plenty of reason to be sceptical.

But I urged myself not to be too hasty. After all, a new, not-avowedly-anti-semitic Director-General was at the helm and they had just decided to engage *me*, and my celebrated strategic and tactical skills, to help them turn the organization around.

Surely that was a sign of keen judgement, if ever there was one?

Chapter 2

Penetration

Hans D'Orville, The Director of Strategic Planning—or rather, to use UNESCO-speak—the Assistant Director-General for Strategic Planning, is a large, corpulent man with a deep booming voice. Originally trained as an economist, he has been diligently working his way up through the UN system for over thirty-five years, and has been in charge of UNESCO's Strategic Planning department for the last decade.

D'Orville and I first encountered each other in the spring of 2010 through a rather curious set of circumstances. The Canadian Ambassador to France had recommended that I meet with Ms Bokova who was then in the midst of putting her new senior management team together. Apparently, the Ambassador and the new Director-General had recently met up at one of the regular schmoozy functions that they are both obliged to participate in, when Ms Bokova had let drop that she was looking to hire someone as the director of strategic oversight or some such thing. The Ambassador had thought that this might be a good fit with my interests and abilities and duly suggested that I contact her office to arrange a meeting to investigate possibilities. Which I promptly did.

But after several months of appointments being made and cancelled owing to her frenetic schedule, I was eventually shunted to the somewhat less peripatetic Mr D'Orville instead, who promptly informed me that the strategic oversight position I was inquiring about simply didn't exist. The notion that the new Director-General of UNESCO would explicitly mention the need to fill a non-existent position to a foreign Ambassador struck me as rather curious to say the least, but under the circumstances there didn't seem terribly much more to say about it. But then D'Orville mentioned the prospect of me starting a fundraising consultancy in the fall. This hadn't exactly been the sort of thing I had in mind, of course, but it seemed—for the very reasons I went on about during the opening chapter—a not unreasonable place to start to see if any mutually satisfying fit could be established. So I agreed.

We met again in mid-September to iron out the details. I was keen to get going sooner rather than later to quickly get a sense of the UNESCO landscape. October might work, I was told, but the beginning of November would be preferred for logistical reasons. Doing my best to appear agreeable, I merrily assented to November 1. We shook hands on that and I was told that I could expect a draft contract to wing its way to me by email in a week or so.

The contract never materialized, despite my regular barrage *I'm sure you're very busy but I was just wondering...* emails to his secretary, which consistently met with resounding silence. By the last week of October, I had ruefully concluded that, having had neither contact nor contract since our brief chat of eight weeks ago, this job was as fundamentally non-existent as the one that had first naively brought me to UNESCO's doors.

Perhaps, I mused philosophically, spontaneous job disappearance was a daily phenomenon in UNESCO-land, with virtual employment popping in and out of the Parisian vacuum like subatomic particles, according to some strange United Nations Uncertainty Principle? Maybe if I hung around outside the main building long enough, I could catch the Director-General job as it flew by and end up running the place—if only for a moment or two before it was someone else's turn.

At any rate, it was surely time to take the bull by the horns and pick up the phone.

When I rang the secretary's number, D'Orville picked up immediately.

"*Yes?*" came an edgy voice, palpably demonstrating its annoyance at being disturbed.

"*Dr D'Orville?*" I queried hesitatingly, confused that he would be answering his secretary's line, but still with the presence of mind to respectfully utilize the academic title that my instincts sensed he would demand.

"*Yes?*"

"*It's Howard Burton. We met some time ago and discussed a consultancy. I was just wondering—*"

"*Oh, yes. Could you start December 1 instead?*'

"*Excuse me?*"

"*It would be best if you could start a bit later, as it happens. I'm doing a lot of travelling this month. Would that be all right?*"

"Well, I suppose so," I agreed. "*But I still haven't received the draft of that contract we had discussed...*"

"*Yes, of course. I'll send it out right away.*"

This was followed by another month of stolid radio silence from his office despite my best efforts to appear eager without overtly harassing. By the time December 1 rolled around, it seemed clear that nothing would be forthcoming and I left my Parisian apartment that morning with a mixture of relief and bemusement, determined to forget about the whole silly business once and for all and spend the day roaming around the Louvre instead. But a few moments later, mulling over my espresso in a local café, I had a sudden change of heart. I pulled out my cell phone and decided to make one more last-ditch effort to assure myself that I had truly exhausted all reasonable options before mentally consigning the entire bizarre experience to the realm of an after-dinner anecdote (*"Have I ever told you how I once almost worked at UNESCO?"*).

I punched in the number and settled in to wait for the secretary's voice mail greeting that I knew so well.

"Yes?" I was met with the same edgy, impatient voice of slightly more than a month ago.

"Dr D'Orville? It's Howard Burton—"

"When are you coming in?"

Suddenly, I found myself on the defensive. When was I coming in? Coming in *where*? *Why*? No contract, no previous appointment, no response to any of my increasingly anxious inquiries about how to proceed, no indication of where I should go or whom I should meet with.

"Um. How about 9:30?"

"Better make it 10. My secretary gets in then."

"OK".

So, no Louvre that day, then, after all.

At 10 am I meet the secretary and am directed towards an HR staffer who promptly hands me a copy of my long-awaited contract, while tersely informing me that I need to sign it back to her later that day. Any irritation associated with this sudden sense of contractual urgency is swiftly dwarfed by a hypothermia alert as I am ushered into my glacial new office (only temporary, I'm informed, given that it is designated for a new hire who will be showing up in a month or so). I spend a few moments forlornly fiddling with the thermostat, desperately trying to convince myself that my frozen fingers are accomplishing something, before settling in behind the desk. The office is equipped with an impressive view of the south end of the École Militaire, but not much else: no working phone or computer or papers of any sort. I spend some time gazing out upon various military equestrian couples refining their dressage before I suddenly remember my contract and unfold it from the coat I am still wearing. Something to do! I rub my gloves together and ponder the legalese for a moment or two under my condensing breath, before opting to take this peculiar little show on the road. There must be a coffee shop or two in the building—now's my chance to locate it and pass a few hours mulling over my contract in relative warmth.

Later that afternoon, D'Orville and I encounter each other for the third time. Waving me towards a couch in his office, he settles himself down on a chair across from me and begins holding forth on his fundraising philosophy.

Fundraising is a notoriously tricky subject, I am informed. It turns out—surprise!—that even obscenely rich people are sadly reluctant to relieve themselves of large chunks of their fortune at the drop of a hat, often stubbornly clinging on to every last dime. Meanwhile, truly productive fundraisers are as rare as hen's teeth: everyone constantly promises the moon and very few seem able to deliver the goods. "*If I had a dollar for everyone who promised me millions of easy money, I wouldn't need fundraisers at all,*" he smirks knowingly, shaking his head.

I smile thinly at this, determined not to start off on the wrong foot and exacerbate a situation which is already quickly careening towards the dangerously unpalatable. After all, his office, at least, was a good deal warmer than mine.

"*The key to doing things differently,*" he continues sententiously, shooting me a searching look to gauge my reaction, "*is to be innovative.*"

I nod my head dumbly, suddenly finding myself close to the breaking point. What in God's name could have possessed me to consider for one moment working at this stodgy fortress of self-importance? Why couldn't I have been smart enough to have headed straight to the Louvre that morning and surrounded myself with an inspiring panoply of master works instead of being held captive by this bombastic UNESCO-vite spouting frothy tautologies of the newness of innovation?

Of course *he'd* be precisely the sort to start droning on about the power of innovation. It's a well-established fact that the least innovative people on the planet are by far the most dogged advocates of the paradigm-changing importance of new ideas. I'll leave it to the psychologists and sociologists to uncover why: Jealousy? Transference? Wishful thinking? Confusion? Who knows? I certainly don't. But it's particularly striking, and not a little amusing, how virtually every pencil-pushing bureaucrat one meets seems to consider himself an authority on the mechanics of innovation and risk-taking. I had seen this particular movie many times before.

So yes: innovation. Certainly. A vital ingredient to success, innovation, particularly when zealously aimed in a decidedly positive direction. For that, too, it must be emphasized, is an integral component of progress. To take but one illustrative example off the top of my head, regularly expecting people to magically show up for their first day of work without any preliminary instruction, a draft contract, or even anything resembling the faintest inklings of a job description is unquestionably an innovative way to conduct HR policy, but hardly the sort of approach that promises to increase one's competitive advantage in the long term. It is a most desirable quality indeed, innovation, but only when pointed towards a meaningful goal.

D'Orville meanwhile, oblivious to my private musings, was plowing steadily ahead.

"*Take the famous Tobin Tax, for example,*" he continued with a sneer, levelling his eyes at me.

"*OK,*" I responded uncertainly, feeling obliged to audibly demonstrate that I am still paying attention.

"*Now that's not innovative at all,*" he cried, banging his fist emphatically on the coffee table separating us. "*It will never work.*"

"*Hmmm,*" I respond again noncommittally, resisting the temptation to point out the obvious structural difference between being innovative and being effective, a distinction that seems curiously lost on the Assistant Director-General for Strategic Planning. Another very bad sign.

"*You're familiar with the Tobin Tax, I presume?*" he presses me, eyes narrowing again.

"*More or less. A tax on financial transactions of something like a tenth of a percent that could result in huge new revenues for government if applied across the board.*"

"*That's the theory. But of course it will never work. Because everyone will simply move their business somewhere else to escape the tax.*"

I shrug my shoulders lightly. I'm not an economist and have never pretended to be one. Generally speaking I don't even like being around economists, as it happens. Still, if forced to wade into these matters it seems fairly obvious even to me that any effective levy on international financial transactions would necessarily entail some sort of broad-based multinational agreement so as to ensure that the tax would be uniformly imposed throughout a wide geographical area. That way, if every major market imposed the tax, the only way to avoid paying it would be to relocate somewhere much less convenient where there would presumably be many other disadvantages that would outweigh its relatively small cost. Whether or not a Tobin Tax could be sustainable at the end of the day is anybody's guess, but I think it's safe to assume that James Tobin, the Nobel Prize-winning economist who first came up with the idea (it's not a *real* Nobel Prize, you understand, just one that some banking guys dreamed up in a rather dismal attempt to try to convince people that economists were engaged in a legitimately scientific activity—but I digress), spent at least some time considering how best to circumvent the obvious concern of a potential flight of capital.

But all of that struck me as flagrantly besides the point anyway.

After all, it's one thing to debate whether or not a Tobin Tax could be sufficiently widespread so as to become an effective revenue-generating device for a spectrum of national governments, quite another thing entirely to make the case that these same governments would feel compelled to hand over even the smallest fraction of such potential revenue to the likes of UNESCO.

Somewhere along the line, all of these tactical speculations would have to give way to a concrete declaration of what precisely UNESCO needed and why. That was the logical starting point to the discussion that oddly hadn't been touched upon yet.

"*How much money are we talking about here?*" I asked gingerly, trying to nudge matters in a more quantifiable direction. "*And for what, exactly?*"

He rose slowly from the couch, all the while examining me closely, evidently having second thoughts on the suitability of someone of my clearly heretical temperament playing any sort of meaningful role within the confines of the illustrious Department of Strategic Planning. Once he had finally made it to his desk, he plunged down to open a lower drawer before surfacing with a heavy glossy tome and various other papers before lumbering back in my direction.

"*Here,*" he announced brutally, dropping the first load on the coffee table in front of me with a thud. "*This is the EFA Global Monitoring Report that demonstrates in detail how we need an extra $16 billion annually to meet our Millennium Development Goals in Education. And here—*" he dropped another fifty papers or so attached with a large paper clip, "*is the summary material for the seminar on innovative financing in education that we held here in mid-September. And here—*" another smaller thud—"*is the report of the committee for innovative financing in education. So you should begin by familiarizing yourself with all of this.*"

I collected the papers and made my way back to my refrigerated office. So we'd start with education, then: the "E" in UNESCO, as it were. As good a place to begin as any.

Chapter 3

The Education Begins

If you suffer from chronic insomnia, you could do worse than have a copy of UNESCO's 2010 EFA (Education for All) Global Monitoring Report by your bedside. More than 500 pages of turbid prose, sanctimonious urgings, accusatory photographs of "the marginalized" and a bevy of detailed figures, tables and charts delineating just how depressingly far we remain from achieving our litany of ambiguously-articulated goals.

For if there's one thing both UNESCO and the international education community seem particularly good at, it's holding conferences and coming up with an imposing array of linguistically-tortured objectives. In the spring of 2000, in what was uniformly hailed as a landmark accomplishment, no less than six separate Education for All (EFA) goals were adopted at UNESCO's World Education Forum in Dakar:

1. Expanding and improving comprehensive early childhood care and education, especially for the most vulnerable and disadvantaged children.
2. Ensuring that by 2015 all children, particularly girls, children in difficult circumstances and those belonging to ethnic minorities, have access to, and complete,

free and compulsory primary education of good quality.
3. Ensuring that the learning of all young people and adults are met through equitable access to appropriate learning and life-skills programs.
4. Achieving a 50 per cent improvement in levels of adult literacy by 2015, especially for women, and equitable access to basic and continuing education for all adults.
5. Eliminating gender disparities in primary and secondary education by 2005, and achieving gender equality in education by 2015, with a focus on ensuring girls' full and equal access to and achievement in basic education of good quality.
6. Improving all aspects of the quality of education and ensuring excellence of all so that recognized and measurable learning outcomes are achieved by all, especially in literacy, numeracy and essential life skills.

Reading through this meandering wish list, I felt a strange sort of nausea sweep over me, the first stirrings of what I would later call "my Texas gun-rack syndrome": the perplexing condition whereby I, a self-proclaimed tolerant, liberal, idealist, committed to the principles of multi-nationalism and international aid and consistently disparaging of the wanton superficiality of our profit-driven, soulless age, suddenly find myself closely identifying with a rifle-bearing, red-state, Republican, convinced that the United Nations is much more the problem rather than the solution and we'd all be far better off if someone would have the good sense to burn the whole damned thing to the ground.

Perhaps it was the stark clash between the boundless self-congratulatory rhetoric and underwhelming outcome that abruptly pushed me over the edge. For the facts of the matter are that somewhat more than a decade ago, high level representatives of 164 nations met for three days in Dakar with the explicit aim of producing tangible goals to alert the world to the importance of developing global educational standards and all they could come up with were six insipid, meandering, flagrantly-redundant statements. Any undergraduate in a donut shop could have produced something considerably more coherent while saving us all a good deal of time and money.

But of course, that wouldn't have been the way to go. International conferences, one repeatedly hears, serve the vital function of "raising awareness of the issues". Fair enough. But it's hard not to feel that the cause of awareness-raising would be much better served if those doing the raising could formulate their points in clear, measurable, meaningful prose rather than tangibly demonstrate how much they themselves are in need of educational enhancement (*Education For All!*). Instead, we get this mess: six incoherent, platitudinous statements that bring the otherwise sympathetic soul to a point of mind-numbing indifference, if not downright Texan gun-toting hostility.

Perhaps you feel, Dear Reader, that I'm getting rather too carried away. Well, perhaps I am. After all, in a world littered with venal, grasping anti-heroes, I grant you that it does seem rather curious to take aim at the relatively few souls who are at least ostensibly trying, in their muddled own way, to make a positive contribution towards the lives of the less fortunate and redressing the global economic balance, however turbid and vacuous their prose might be. But there's the rub: it's precisely *because* of my deep-seated convictions of the over-arching importance of the cause that it's so infuriating to find it left to a banal cabal of, inept, self-congratulatory buffoons.

An exaggeration? Maybe. But let's take a moment to invoke our well-honed critical thinking skills to rigorously examine the EFA objectives that were the culmination of this landmark three-day meeting. Goal number 3 sententiously informs us that young people and adults should learn by appropriate programs—hardly the sort of thing that needs to be mentioned, I should think. Goal number 5, meanwhile, specifically invokes the cause of gender equality. Given that the systematic discrimination against women and girls in the developing world is a leading factor in stunting both educational advance and—relatedly—economic prosperity, this is highly appropriate. So mention it loudly and proudly by all means. But then why add in goal number 2 that the avowed aim is not simply to ensure that *all* children should be in school by 2015, but *also* "*particularly girls, children in difficult circumstances* (whatever that means) *and those belonging to ethnic minorities*". Doesn't the word "all" unequivocally imply girls, boys, members of ethnic minorities, members of ethnic majorities, right-handed, left-handed, disabled, non-handicapped and so forth? Don't they *realize* that by continually backing away from the universality of their cause ("especially for the most vulnerable and disadvantaged", "especially for women") irreparable harm is done to the entire essential *concept* of universality. That is a big deal, after all. They don't call it the "Universal Declaration of Human Rights, Particularly For Vulnerable Minorities". It's just the Universal Declaration of Human Rights. Period. Because it applies to *everyone equally.* Indeed, that's the whole *point* of the thing.

Then there is the issue of the quality of the educational experience, the topic of Goal 6, but lurchingly referred to in Goals 2 and 5 for good measure. In addition to ensuring that children physically attend school, it seems that we should also pay some attention to what they are actually doing when they get there. Great idea. But despite the disturbingly Orwellian prose that urges us to ensure "excellence of all", I'm guessing that educational standards will likely vary considerably from country to country and that a more reasonable option under the circumstances would be to explicitly invoke quality as a desirable goal to be attained and eventually objectively assessed, along with the likes of access, affordability and adult literacy rates.

Again, it might well be suggested that I am being far too hard on the educational boffins here. After all, producing succinct proposals from a multinational setting is bound to be difficult: if a camel is a horse designed by committee, then one can only imagine how hideous the beast resulting from the confluence of 164 separate national interests might be.

But it's worth bearing in mind the context here. We're not talking about the vastly more complicated challenges of *actually achieving* success or even the considerably easier but still awfully thorny issue of deciding how, precisely, to measure it. And nobody is questioning the integrity or even feasibility of the avowed priorities (full access to primary education, expanding early childhood education, improving adult literacy rates). All we're trying to do here is simply clearly and intelligibly delineate what these priorities are. And if we can't even do that, then what's the point in continuing at all?

Quite frankly, it's really not all that difficult. It's worth pondering what our donut-dunking undergraduate might come up with after half an hour or so of reflection. Doubtless something along the rather self-evident lines of:

1. Ensure all children have access to free, compulsory quality education by 2015
2. Actively promote gender equality throughout all levels of education
3. Expand and improve early childhood education programs
4. Achieve a 50% improvement in adult literacy by 2015

Behold four clear, pithy objectives that at least make for a reasonable starting point.

Of course, there will be a corresponding need for considerably more clarification when one turns one's attention to the nitty-gritty of performing explicit evaluations. How does one precisely measure "basic educational quality" or "the level of early childhood education programs"? What are the standards? What are the specific means invoked to promote gender equality and how can one explicitly evaluate the impact that such programs are having? How should literacy be measured? And so on.

All of this will naturally be difficult. But without first having clear, transparent objectives to move towards, it will simply be impossible.

Thankfully, at some level even the powers that be at the United Nations seemed to recognize this, which is why the 8 Millennium Development Goals (MDG), created in September 2000 some five months after the Dakar conference, at least enjoy a measure of brevity and concision that vastly improves upon UNESCO's rambling EFA disaster.

Two of the eight MDG apply specifically to education. The second MDG aims *"to achieve universal primary education"* with an accompanying target to *"ensure that by 2015 children every-*

where, boys and girls alike, will be able to complete a full course of primary schooling". One can quibble about the fact that UN goal-writers still seem curiously averse to employing standard universal quantifiers (why not simply "all children"? Why do we once more need to be reminded that children come in both girl and boy flavours?), or wonder meditatively what precisely the difference might be between being *able to* complete a full course of primary schooling and *actually completing* one, but these are relatively minor points. Here, at least we have a fairly concrete goal that one could reasonably imagine being both comprehended by and resonating with someone not directly employed by a UN agency.

The third MDG, which aspires to *"promote gender equality and empower women"* while invoking a target of *"eliminating gender disparity in primary and secondary education preferably by 2005, and in all levels of education by 2015"*, is also an improvement upon the previous EFA Goal number 5, having at least pared away the Einsteinian insight that eliminating the gender disparity in education would necessarily result in girls having full and equal access to schools. So this too is much better, but still manages to intellectually offend by the injudicious decision to insert the word "preferably" into the proceedings. Was the target to *actually* eliminate gender disparity in primary and secondary education by 2005, or not?

It is unquestionably *preferable* to eliminate it by 2005. For that matter, it is *preferable* to have eliminated it by 2001. It is doubtless far more *preferable* still for it to have never existed at all. But then, doing what is preferable is surely what the Millennium Development Goals are all about. That is precisely why they exist, in fact, to inspire us to move from the status quo to a better world. Moving to a *less* preferable world would be something else entirely, and presumably involve invocation of the Millennium Lurking Fears (MLF) or Millennium Retrograde Steps (MRS), which is well outside of our present purview (although quite possibly the subject of at least one other UN-sponsored conference).

Of course, I know what they *meant*. Back in 2000, the thinking was that we should *really try* to eliminate primary and secondary school gender inequality by 2005, but that it was *probably* unrealistic to do in five years and a safer bet would be to shoot for 2015 in tandem with all other levels of education. But then why not just say 2015 in the first place? That way you start off with a much more realistic target, and if you somehow manage to achieve it by 2005 you can proudly trumpet your magnificent achievement of having accomplished a key goal ten years earlier than hoped. We're hardly talking high-level strategy here; and once again, it's difficult to become optimistic about the future of global education when its self-proclaimed leaders seem to have such a tenuous grasp of either basic human psychology or the simple meaning of words.

It turns out, unsurprisingly to anyone even remotely acquainted with the deeply entrenched societal gender bias of so many UN Member States, that 2015 is as hopelessly optimistic a date for the universal educational parity of girls as 2005 was. More generally still, the prospect of substantially meeting any of the EFA or educational MDG 2015 goals is virtually non-existent. But thanks to the authoritatively weighty 2010 EFA Global Monitoring Report, we at least get to learn what is responsible for this sorry state of affairs: not enough money.

Sandwiched by the usual profusion of bombastic near-tautologies (*"poor quality education in childhood has a major bearing on adult illiteracy"; "the core purpose of education* is to ensure that children acquire the skills that shape their future life chances") lies the over-arching conclusion that the principal impediment to attainment of the six EFA goals is insufficient funding.

"*Ten years after Dakar*", the report laments, "*finance remains a major barrier to Education for All.*" A detailed EFA cost analysis has been undertaken, we are informed, before finding ourselves rather cavalierly informed that an additional injection of "*around $16 billion annually*" is required to meet the EFA goals in basic education.

Somewhat refreshingly, the report specifically mentions two distinct avenues towards finding the missing resources to inject into the system: national governments and donor countries. "*National governments can raise a substantial share of the additional resources needed,*" we are told, before being confronted with a starkly illuminating bar graph contrasting, on a country-by-country basis throughout the developing world, the 2007 level of educational spending to what would be required to meet the 2015 goals.

Curiously enough, however, by the time we have moved from the full 500-page report to the considerably more accessible 38-page summary that presumably enjoys a vastly wider readership (as far as these things go), virtually all talk of any appreciable role played by national governments—suddenly, revealingly, called "recipient governments"—has disappeared, and the onus is now squarely on the shoulders of the rich world's donor countries who are irresponsibly falling short on their commitments. Of course at this point, well above and beyond the usual challenges of finding a way to substantially increase international aid, most of the developed world is currently preoccupied with the global economic crisis that is raging around them, and in no mood to treat any of this as a priority.

So this is what all of our charts and graphs boil down to: UNESCO claims that we need $16 billion more each year for basic education of the world's poor. Poor countries won't pay. Rich countries won't pay. So we need to find substantial new monies somewhere else: the so-called: "innovative financing" scenario that Dr D'Orville was adamantly insisting upon earlier.

Except that when you get right down to it, there are really not all that many actually innovative approaches to innovative financing. One can either introduce a broad-based tax on goods and/or services (financial transactions, mobile phone bills, lotteries, petrol, what have you), or somehow induce new philanthropists, foundations or corporate partners to step up to the plate. The more technically minded are often prone to mooting more sophisticated economic vehicles (currency education bonds, diaspora bonds and so forth), but these tend to be nothing more than glorified mechanisms for additional national-government support or specifically directed philanthropy (from ex-patriots). A somewhat more intriguing idea is the debt for education swap, whereby a creditor nation agrees to cancel a long-standing debt by specifically redirecting the funds towards specific development programs. But at the end of the day, this is effectively indistinguishable from old-fashioned foreign aid.

The impressively large elephant on the table, of course, is that there is widespread—and eminently justified—scepticism concerning what sort of effect any putative new monies might actually have on the status quo even if they could somehow be found. The governments of Chad, Equatorial Guinea and Nigeria, to name but three of many obvious examples, could easily direct far more resources towards the basic education of their citizenry today should they so desire. But they don't. Which rather ostensibly begs the question: Why not? And, relatedly, how certain could we be that the fruits of any successful "innovative financing" initiative might actually find their way towards their intended recipients? Answer: not very. Which in turn makes it naturally all the less likely for, say, well-motivated ex-patriot Nigerians to participate in any new fundraising scheme, however ostensibly "innovative".

But these aren't the sort of questions that our bureaucratic financing experts seem terribly preoccupied with. A quick perusal of the 2010 report by the Writing Committee commissioned by the Task Force on Innovative Financing for Education created by the Leading Group on Innovative Financing for Development (see how quickly the eyes mist over?), reveals the standard laundry list of theoretically plausible fundraising mechanisms followed by solemn recommendations to form new expert committees to hold more roundtables to seriously study the matter.

By this point I was dangerously close to merrily consigning all of this sanctimonious claptrap to the flames and walking away from the whole sordid business entirely. After all, what possible constructive role might someone like me play in this bizzaro universe where UNESCOvites spend their days sitting around ever-increasing round tables, wringing their hands despairingly at the boundless selfishness of the human condition that cruelly prevents them from funnelling arbitrarily huge sums of money towards corrupt despots in the hope that some of it might magically filter down to their bevy of incoherent, nebulously defined programs?

My sole approach to fundraising has always been to directly approach would-be partners and enthusiastically hold forth on the precise sort of concrete, transformative programs putative joint efforts might happily produce. That clearly couldn't happen here. Aside altogether from the obvious fact that I wasn't personally responsible for any particular UNESCO initiatives and thus could hardly give personal assurances of taking ownership of any particular project, there was the rather more sizeable constraint that I had yet to discover even one existing program that I had any genuine confidence in. Put another way, if I were a billionaire anxious to make an impact in educational philanthropy, I would be hard-pressed to envision myself giving a nickel to UNESCO given all that I had read and seen to date. Convincing others of the merits of a cause can often be difficult, but it's nigh-on impossible if you can't even convince yourself.

I tossed the innovating finance report on top of the growing mound of paper on my desk and trudged upstairs to the communal coffee area to grab an espresso, clear my head, and try to calmly ponder whether it made any sense at all for me to stay one more minute in this god-forsaken den of bureaucratic iniquity. Making my way towards a solitary unoccupied table, I gazed out towards the Eiffel Tower rising through the Parisian mist and felt a languorous calm descend upon me, my inner Texan summarily beating a swift retreat at the sight of the iconic metal structure reigning over the Champ de Mars. UNESCO may well be the most dysfunctional member of the entire United Nations apparatus (which is no mean accomplishment), but you sure can't beat its location.

"*Hi there*," a voice greeted me, punctuating my francophilic musings.

I looked up to see Jacques, my sole UNESCO social contact to date: a wry, engaging Moroccan who toiled away in another corner of the Department of Strategic Planning, currently occupying himself translating the meeting notes of the recent UNESCO seminar on Innovative Financing for Education from French to English. Or perhaps it was English to French. Not that it made much difference: Jacques could certainly do either with equal ease. And it was hardly clear who would bother taking the time to muddle through it in any language.

"*Hey*," I replied, motioning for him to take a seat.

"*Settling in?*"

"*Not really. Frankly, I'm not sure I can stand too much more of this.*"

He chuckled. "*Well, that didn't take too long. What is this, day two?*"

"*Three*," I corrected him proudly. "*Only weenies seriously contemplate bailing after the second day.*"

Jacques was by far the most captivating thing about UNESCO that I had encountered so far. Here was an astute, educated, well-travelled, individual who seemed painfully aware of the vast majority of UNESCO's limitless shortcomings, and yet still elected to work there. He had dropped in to my office to introduce himself a few hours after my arrival and, after a few tentative, preliminary comments, we quickly embarked upon a long and illuminating discussion about UNESCO and its global impact. Or lack thereof.

I discovered him to be an intriguing combination of passionate idealism and rampaging cynicism, the likes of which I had never before seen. One moment he would be excitedly carrying on about how UNESCO could potentially be transformed into a gleaming beacon of enlightenment for humanity, leveraging the bona fide success of the World Heritage Sites program to specifically increase environmental awareness, partnering with motivated NGOs to make a strong international stand against human rights violations and debilitating corruption, effectively promoting tolerance and understanding through global educational exchanges and so forth.

And then he would stop in mid-sentence as if suddenly comprehending the enormous disconnect between what could conceivably be and what actually is, abruptly switching gears to bitingly educate me on UNESCO's rigid silo-mentality that ensures the highest possible levels of inter-organizational tribalism and short-sightedness, its penchant for consistently recruiting the most benighted and ineffectual individuals possible for senior positions through the rigid invocation of an HR policy resolutely centred on geographical parity, intellectual timidity and a strict adherence to the status quo, its congenital incapacity to ever look at itself the slightest bit objectively or make meaningful comparisons with other similarly-oriented institutions and so on.

One minute I'd be speaking with Gandhi. Then, suddenly, I'd be confronted with Christopher Hitchens. It could sometimes be perplexing, but was invariably interesting.

That afternoon in the coffee lounge, his cynical side took the lead, as he began to probe how I came to find myself in the Department of Strategic Planning in the first place.

"So you didn't know D'Orville or anyone else before you came here?"

"No."

"And you ran a physics institute, right?"

"Yup. Public-private partnership. The magical three p's."

"Hmmm. Interesting."

"At times."

It was then that I took the opportunity to launch into a detailed articulation of my fundraising philosophy: how I viewed it as inextricably related to strategic development and how it was so often the case that institutions, through the dead weight of ever-accumulating inertia, found themselves focused exclusively on revenue-generating tactics rather than the much more significant questions of *what* was actually needed and *why*.

Perhaps, I suggested hopefully, the Department of Strategic Planning had recognized this inherent bureaucratic weakness and felt the need to shake things up by bringing in an experienced, outside observer to provide some external, objective advice.

"*It's possible*," he agreed dubiously, making a pointed effort to not be completely dismissive of my idealistic delusions on my very first week. "*Or perhaps they just thought that you might be able to give them some easy cash through your connections to rich people. You do know some rich people, right? What about the guy who funded your institute? The BlackBerry guy. Isn't he loaded?*"

"*Sure. But he wouldn't give a dime to UNESCO or any other large, bureaucratic place. That's hardly his style. Why do you think he wanted to start something of his own?*"

"*Makes sense*," he smiled.

I nodded my head before looking at him closely.

"*So you think that's why D'Orville hired me? Because I might be able to bring in some easy money?*"

He shrugged his shoulders.

"*I don't know. Probably. Maybe. You said that this all started by some meeting with the Canadian Ambassador and the DG, right? Perhaps D'Orville thought you could somehow get Canada to up their UNESCO contribution. Or maybe it was all the DG's idea and he was just following orders. Who knows?"* He shrugged his shoulders again before musing, mostly to himself. *"It **is** a bit odd, though, that you're not working directly with the Department's development section."*

"Strategic Planning has its own fundraising unit?" This was news to me.

"Yeah. Three or four people. They mostly do technical stuff: formal relations with foundations and banks and that sort of stuff. Nothing terribly proactive. They haven't exactly been bringing home the bacon."

"So you think that D'Orville wants to shake things up there?"

He shrugged again. *"I don't know. Could be. It's hard to tell. Maybe it's just another typical UNESCO screw-up. Nobody talks to each other here anyway. The amount of pointless duplication in this place will simply blow your mind—you'll find that out soon enough,"* he added with another indulgent smile, *"if you stick around for a bit."*

"You're not making it very appealing, I have to say."

"No," he admitted. *"Probably not. Sorry about that."*

He paused for a bit before continuing in a somewhat more thoughtful tone, his Gandhi-side once again suddenly rearing its temperate head.

"I shouldn't really be so negative. It has its good points, you know, UNESCO."

"Such as?"

*"Well, it's a good **idea**, at least. That means quite a lot, actually."*

"Does it? Marxism was a good idea too. And then Lenin came along. And then Stalin."

"And you shouldn't forget about all the people out there in the field," he continued, ignoring this barb. *"The ones that do the heavy lifting and actually have a real impact. UNESCO is not just about what happens at headquarters in Paris, you know. There are lots of people doing very good work in very challenging environments with very few resources."*

"Without a doubt," I agreed, not having any direct understanding of whether or not this was true, but certainly quite willing to take it on faith. *"But doesn't that make it all the more important to try to change how things are done here in Paris? For their sake? So that what they're doing might actually make a difference?"*

He slapped me on the back amicably as he got up from his chair.

"Very well said. You know, you should really stick around for a bit. This place needs you."

*"You mean **you** need me. I think you're just lonely."*

"Maybe so," he smiled again. *"Well, somebody needs you, anyway."*

Chapter 4

2+3=8 and Other Revelations

Thus inspired I trudged back to my office and picked up the innovative financing report on my desk. Perhaps I was being too hasty. Maybe there was something worthwhile in there after all?

The title, it must be admitted, had put me off a bit: *"2+3=8—Innovating in Financing Education"*. The idea behind such an iconoclastic approach to basic arithmetic comes from Jordan's Queen Rania Al Abdullah, who once publicly averred that the second and third millennium development goals—primary education and gender equality—have a preeminent importance for achievement of all eight, thereby illustrating *"the only case I know where 2+3=8"*.

This is, I think, a dubious argument. However meritorious the causes of the second and third millennium development goals might be, it's awfully difficult to convince oneself that achievement of gender parity and full primary schooling by 2015 will simultaneously result in, say, the halving of 1990 extreme poverty levels. Indeed, a strong argument could be made in the other direction, positing that education and gender equality programs can only be truly effective once vast numbers of the students involved are no longer starving.

But this is probably quibbling. Obviously both education and poverty reduction must be vigorously pursued in tandem; and there can be little doubt that an integral aspect of a country's long-term success and economic self-reliance lies in its ability to educate its citizenry.

Moreover, notwithstanding her decidedly lightweight educational qualifications as socialite and Royal consort, her embarrassing tweets during the beginning of the Arab Spring urging Tunisians to meekly seek calm and stability rather than overthrow their oppressive dictator, or the various allegations of corruption and the abuse of the public treasury for her own personal ends that have been lobbed against her, it cannot be denied that the Queen has been a tireless and sometimes influential advocate for both women's rights and universal education.

So perhaps 2+3 doesn't really equal 8 after all, at least in the sense that Queen Rania avers. But all that is simply window dressing. Surely the point is to highlight the fundamental importance of universal education and consider how innovative approaches to financing might somehow result in this worthy goal becoming a reality.

A closer look at the report reveals some encouraging remarks I had missed in my original exploratory run-through. Point 7 of the 20 points made in the meandering Executive Summary explicitly states: *"A major argument in favour of introducing innovative financing mechanisms lies indeed in the belief that these new sources of funding will not only generate additional funds, but will also contribute to generate a virtuous circle of change in the education sector by increasing the mobilization of domestic resources, improving aid effectiveness, fostering innovation and improving performance."*

Aha! Finally! Innovative financing is much more than simply finding some mechanism to extract more money for our nebulous cause. It is all about creating a "virtuous circle of change" so that the money raised will end up being vastly more effective and ensuring that measurable success can and will occur! Hooray!

True, the report also explicitly mentions that innovative financing should be "more stable and predictable" than even traditional aid, which seems to run strongly contrary to the whole idea of tying assistance explicitly to performance (if the funds are guaranteed, how on earth can one reasonably expect that ineffective programs will be shut down?). But still, there seemed reason to hope. Virtuous change was in the air!

I delved into the report with renewed optimism. Section 2.3, I discovered happily, was wholly devoted to this celebrated virtuous circle of change.

"Along with the needs of the sector, another essential argument in favour of introducing innovative financing benefiting education lies in the belief that these new sources of funding could generate a virtuous circle of change, improving the impact and the cost-effectiveness of further investments."

Alright. There's the recap to set things up. On to some concrete details now!

"Stepping up the culture of innovation and risk taking in the education sector will help to achieve much better results even with the finances currently available."

Hang on a minute. Didn't they already say that?

"This joining of innovative financing with financing for innovation has the potential to create a virtuous circle of change that could break through the deadlock currently restricting reform in the global education sector."

Enough with the circular platitudes already! Give me something I can get my hands on!

"Innovative financing should thus take into account the necessity to further promote innovation within the sector, in both fundraising and aid delivery."

And so it goes. A few more serenely redundant paragraphs later, I encounter a circular flow-chart where rounded arrows flow clockwise from "innovative finance" to "improved results" to "more support for education" before being irrevocably led back to "innovative finance". There it is, in black and white and presumably all primed and ready for a bedazzling PowerPoint Presentation: A VIRTUOUS CIRCLE! Who'd have thunk it?

At long last, however, there is actually an *example*, of sorts: it turns out that some clever Washington think tank has developed the notion whereby prospective donors might first demand that specific targets are met before actually giving their money, rather than watching helplessly on the sidelines as the funds get squandered, which seems to be the usual scenario.

This being the aid community, a fancy acronym is deemed necessary to describe this ground-breaking scenario; hence COD—Cash on Delivery. The key features of this new innovation, we are told breathlessly, lie in the unique focus *"on outcomes rather than inputs"*. That is, the innovative power of COD lies in the fact that the recipient actually has to *first* clearly define what needs to be done and *then* go ahead and accomplish it, as opposed to the standard Cash on Demand sort of affair (the previous COD, one imagines acronymoniously), where donors simply pony up untraceable funds once they are told to. This new-style COD, we are informed, *"has received several expressions of interest from developing countries and some donors."*

What to do when confronted with this litany of sanctimonious drivel posing as cutting-edge thinking? I despairingly flung the report back onto my desk and decided that perhaps it was time for another approach. Given that two of the twelve authors of the report were employed by UNESCO, perhaps it would be easier just to drop by and see them directly.

It wasn't, I hasten to add, terribly clear what I should see them ***about***, mind you. In fact, the deeper I delved into this curious world of redundant hand-wringing, the less convinced I was that I had anything significantly to contribute at all, so unbreachable seemed to be the gulf between my world-view and that of the Professional Innovative Financiers. Still, I presumably should be doing ***something***.

After all, the head of Strategic Planning who had engaged my services had specifically suggested that I start off by innovatively finding the missing $16 billion for education. At considerable mental anguish, I had waded carefully through a voluminously vapid report on the matter and had isolated a good number of ambiguities and obvious points of concern. Under the circumstances, it seemed the least I could do to take the elevator down a couple of floors to try to gently bring my objective perspective to the attention of the authors. It was becoming increasingly obvious to me that I wasn't going to be UNESCO's innovative financing messiah, but perhaps I could at least nudge the community a tiny way towards writing better reports.

Such, at least, was my hope. I arranged a meeting with the junior UNESCOvite author, and spent several hours in advance carefully preparing tactfully phrased suggestions that would combine a critical spirit with a general sense of respectfulness for the effort that the Writing Group of the Task Force of the Leading Group had already undertaken.

I naturally opted to focus my attention on tactical details rather than highlighting the larger strategic issues at play (such as what on earth they were realistically hoping to accomplish by all of this). There was no point in starting a discussion waving a red flag in someone's face and casting aspersions on the whole process painstakingly begun a decade or so before I was even a gleam in UNESCO's eye: I was the outsider, simply curious to clarify a few matters and perhaps make the odd benign recommendation or two based upon my experiences. This was hardly the time to question the amounts required or to voice suggestions as to how to ensure that any additional aid might actually be more effective. No, I would swallow hard, act as non-confrontationally as possible, and simply limit any suggestions that I might respectfully make to the internal coherence of the report combined with politely inquiring how one might practically move forwards towards achieving the Herculean financial task before us.

Well, that was the plan. But despite my best efforts, we didn't seem able to get off on the right foot.

I introduced myself, giving a brief history of my background before mentioning that I had recently started a consultancy with the Department of Strategic Planning and was initially encouraged to focus on the matter of innovative financing for education and the celebrated $16 billion challenge.

This, oddly, went beyond the pale.

"*That's just like Hans!*" she fumed bitterly.

"*Excuse me?*"

"*He goes ahead and hires a fundraiser and doesn't even tell us. Education is not the fief of Strategic Planning, you know. This doesn't have anything to do with them!*"

I looked down at my feet guiltily. I had been so distracted by the utter preposterousness of the entire exercise, I had completely neglected the possibility that I might be stumbling onto yet another UNESCO turf war.

"*I just have a few questions about the report...*" I began meekly.

"I don't have much time," she snapped.

"*Well, OK,*" I replied, rapidly unfurling my carefully prepared notes. "*I'll try to make it quick.*"

"*Actually,*" she snapped again, glancing down at my four pages of comments. "*I really don't have any time at all today.*"

This clearly wasn't working. Desperate times call for desperate measures. I suddenly had no recourse than to be completely honest with her.

"*Look, I see that you're upset. And honestly, I don't really blame you. You go to six months' worth of meetings and spend all this time working with eleven other people on drafting a report and then once you're finally done some new, unknown guy suddenly shows up from Strategic Planning anxious to harass you with a bunch of detailed questions. I get it. You don't need the aggravation.*"

She looked at me blankly, clearly unsure of how to respond to this deluge of forthrightness.

"*But look at it from my perspective,*" I continued, pressing my advantage created by her momentary bemusement. "*Strategic Planning hires me because—or so I naively assume—I possess some insight and experience at developing effective and impactful strategic partnerships. I'm told to first turn my attention towards the issue of innovative financing for education and am given your report. So I spend some time reading it over closely and, I have to admit, I find it pretty underwhelming. This is so clearly not my world. But since this is where I seem to find myself at present, I figured I might as well make an effort to give you some sort of feedback and participate in some concrete way."*

"So how about this: let's both pretend for half an hour or so. I'll pretend that the Writing Committee to the Task Force on Innovative Financing for Education might conceivably be the slightest bit interested in my objective assessment of their report complete with some tangible suggestions on how to go forwards, while you pretend that I actually believe that any of this will wind up making one jot of difference to the cause of universal primary education."

It just slipped out. I had meant to be frank, but not quite **that** frank. Obviously, I had overstepped my bounds. Somewhat sheepishly, I got up to leave.

But, curiously enough, she motioned for me to sit back down and made her way purposely over to close the office door before settling back behind her desk.

"The report is shit," she declared after another brief pause, folding her hands together. *"They made us come out with it far too quickly, before we were ready."*

This was unexpected, to put it mildly. I waited quietly for her to continue, resisting the temptation to ask who the dreaded "they" were. It hardly mattered really.

"So alright," she continued after another few seconds of silence. *"Let's pretend, like you suggested. Tell me what your thoughts are. Why not? It's not like I've got anything better to do anyway."*

I wound up spending the better part of the afternoon in her office, candidly discussing the educational realities in the developing world and jointly imagining what sort of effective strategies might be created to concretely improve the situation. So, not withstanding the rocky beginning, it turned out to be an interesting meeting after all, and highly educational in its own way.

The meeting also served to give me further confirmation of my growing suspicions that, notwithstanding their unceasing chest-pounding and fervent declarations of international parsimony, UNESCO was, in fact, increasingly becoming a relatively minor force in global education, falling further and further behind the World Bank and a host of other international foundations and philanthropic organizations. Even when viewed through the rather dubiously efficacious filter of UN agencies, UNESCO's educational efforts had long been eclipsed by UNICEF. In short, in the increasingly cluttered space of international basic education, it was indeed very hard to see *what*, if anything, UNESCO actually did or, put more starkly, what would happen to the cause of global education if it should suddenly cease to exist.

On the other hand, one of the most intriguing educational success stories of late was to be found just on the other side of the Seine at the Organization for Economic Cooperation and Development in the stately 16th arrondissement. To a UNESCOvite, the OECD is roughly equivalent to a Capitalistic Death Star, menacing the planet with its pernicious doctrines of untrammelled market liberalism, the nerve centre of economic imperialism replete with armies of glib, Armani-suited analysts, all ruthlessly determined to ride roughshod over the world's downtrodden in their unceasing promotion of the immoral interests of their unsakably rich members.

For in stark contrast to the shining multicultural universality that pervades the corridors of all UN agencies, the OECD is an unabashedly private club of the planet's 34 wealthiest nations, a wood-panelled country club dedicated to the single cause of perpetuating their oppressive economic hegemony over the world's less fortunate by whatever means possible.

How interesting, then, that this same sinister OECD has found a way to evolve towards becoming a globally dominant player in the field of education: providing a wealth of clear, comprehensive, comparative data for all aspects of the educational experience, from early childhood development programs through to adult education and retraining, objectively searching for best practices while rigorously incorporating key features such as gender bias, teacher training and concrete evaluation of the link between educational infrastructure and economic growth.

In 2000, the OECD established PISA (Program for International Student Assessment) as a means of rigorously evaluating and contrasting educational standards across a wide spectrum of countries, testing a broad mix of 15 year old students on problem solving, mathematics, reading and scientific literacy. The data are thoroughly analyzed and publicly released with explicit country by country rankings every three years, an event which has garnered increasing media attention and justified policy influence for the OECD's bourgeoning Education Directorate. In slightly more than a decade, PISA has become nothing less than the gold standard of objective educational record throughout much of the world, with 75 countries opting to participate in 2010, almost double the amount who were involved a mere ten years earlier.

So it is that the OECD, which owes its institutional origins to the need to create a centralized disbursement vehicle for Marshall Plan funding in 1947, has managed to quickly establish itself as an international educational leader.

Meanwhile UNESCO—which, as the *United Nations Agency for Education, Science and Culture* one might well have thought would be an ideal place to develop a set of broad-based scientifically-rigorous metrics of educational indicators, has become increasingly marginalized and irrelevant, reduced to publishing cavilling lamentations of its own increasing ineffectiveness and convening meetings focused on how fantastic sums of money might somehow magically be purveyed from the resolutely callous international community.

But I had had more than enough of trying to wade through any more of such self-serving protestations. Fresh from my surprisingly revealing conversation with the UNESCOvite report writer, I was more convinced than ever that the only way for me to make any sort of headway on understanding the lay of the land would be through direct, candid one-on-one encounters with experienced members of the development community. At the very least it would certainly be more entertaining.

I picked up the list of participants for the recent UNESCO seminar on Innovative Financing for Education and noticed the presence of a certain Jon Lomøy, Director of the OECD's Development Co-operation Directorate. This was a surprise: I didn't even know that the OECD *had* a Development Co-operation Directorate. This hardly seemed the sort of thing that a den of soulless, neoliberal oppressors would bother spending much time on. A clever ruse, perhaps? Sheep's clothing (did Giorgio Armani make sheepwear?) for the insatiable wolf?

I asked Jacques about it several days later, but he wasn't terribly forthcoming. *"Ah yes,"* he sniffed, *"the OECD. They do some good stuff."* In fact, it turned out that he had once collaborated closely with someone over there and had even been offered a position afterwards. But he turned them down: something about a clash of values.

I decided not to press him further, unsure whether or not this was a reflection of some deep-seated well-spring of anti-OECD sentiment or merely the consequence of randomly catching him at the apex of his oscillating UNESCO biorhythm. In any event, it didn't much matter at that point: by then I had already procured a meeting with Mr Lomøy.

It turns out that I didn't get the full OECD Star Wars experience that day after all, since the Development Co-operation Directorate is located well away from their main headquarters in a separate building in Boulogne-Billancourt. But that didn't seem to make much difference to me.

I spent close to two hours with Jon Lomøy, basking in one of those delightfully rare, wide-ranging, intellectually uplifting discussions when the minutes fly by so quickly that a sudden confused glance down at your watch only convinces you that your battery clearly needs to be replaced.

A soft-spoken Norwegian, Lomøy had held an intriguing mix of influential positions in both government and the development community before coming over to head up the OECD's Development Directorate: Director of the Africa Department at the Norwegian Agency for Development Co-operation; Deputy Director-General for Africa, Asia, Latin American and the Middle East in the Department of Foreign Affairs, Ambassador to Zambia, Ambassador to Tanzania. And so on. The guy had definitely seen a thing or two.

I had ostensibly come there, of course, to discuss innovative financing for education, the daunting $16 billion challenge, and his inside insights on the recent UNESCO seminar. But several minutes into our conversation, it was clear that neither one of us had much appetite for any of that bureaucratic grandiosity. While he was far too tactful a fellow to cast aspersions in any definite directions, it was obvious to me that Lomøy was much too wise and had seen far too much to concern himself with UNESCO's fanciful notions of self-importance. As a good citizen of the development community, he had presumably felt obliged to attend the seminar, doubtless feeling that a quick jaunt across the Seine to the 7th arrondissement was not such a high price to pay to tangibly demonstrate some good faith towards harmonious inter-institutional relations. Perhaps he had the good sense to combine the whole dreary business with a rewarding trip to the Musée Rodin or steal an hour or two browsing in one of the many captivating left-bank bookstores. This was Paris, after all. It wasn't like he had to fly to Winnipeg.

And so we talked about other matters instead. I asked him general questions about what he thought worked and didn't work in development. I wondered aloud on whether sub-Saharan Africa would ever be capable of making the sorts of impressively rapid strides in economic and social standards that countries like Singapore and Korea and Taiwan had undergone; and if not, why not? I queried him on his candid views of the international aid community at large, his overall level of optimism or pessimism regarding the case of international development in both the near and long term and his concrete suggestions for how, generally speaking, it might best be improved.

There was certainly no shortage of things to ask. Here was, after all, someone who had spent considerable time on the ground negotiating deals, initiating programs, and working with the local citizenry before returning to more effectively reorganize departments back home. He had many stories to tell. And the more insightful and revealing his responses, the more motivated I became to ask follow-up questions.

By the time I had exhausted even his considerable patience and reluctantly found myself once again outside of the shiny new building in Boulogne-Billancourt, I was feeling suffused with the warm glow of unbridled optimism. I had just had a substantive, stimulating conversation with an astute, knowledgeable, international aid specialist clearly capable of successfully balancing idealism and realism after a long and varied career. Obviously such people existed. Who's to say that there aren't more of them? Who's to say, in fact, that there aren't even one or two rattling around somewhere within the otherwise bleak corridors of UNESCO? Stranger things have happened (not too many, however, it must be admitted). It's even logically possible that someone who fits that description might be found within the Bureau of Strategic Planning.

After all, my recent experiences at the OECD were concrete proof that large, potentially stifling bureaucracies can somehow manage to harbour the occasional shining point of light. The Organization for Economic Cooperation and Development, it is very much worth bearing in mind, should hardly be viewed as the poster boy of creative dynamism or innovative thinking: all those suits proudly prancing around with their spanking new competitiveness models and their highlighted Excel spreadsheets: where had all that really got them? Where were all the bold OECD announcements alerting the world to

American's festering subprime fiasco well before the Lehman Brothers collapse in 2008? Where were the 2009 clarion calls for action to counter the Euro crisis that burst onto the scene in 2010? Nowhere. You might think that with all their chest-thumping analytical firepower, the Organization for Economic Development and Cooperation could have played some sort of non-trivial role in enabling its Member States to avoid, or at least substantially minimize, the most powerful financial storm they would face since the institution's founding. But no. When all's said and done, most of these guys are just accountants, economists and bureaucrats. Even their hindsight isn't always 20:20, let alone their foresight.

So, I chastized myself, let's not get too carried away here. By hiring some good people, emphasizing the importance of transparent, objective data analysis and correspondingly creating some carefully thought out programs such as PISA, the OECD had managed to firmly establish themselves as a credible international leader in many aspects of international education and development. Good for them. But it has hardly cornered the market on good ideas or best practices. And if they seem so relatively progressive and dynamic, that might well be more of a commentary on the current institutional ossification of places like UNESCO than anything else.

And perhaps, I mused idealistically, all hope was not lost. Perhaps even UNESCO contained a few diamonds in the rough: frustrated pockets of potential just waiting for a suitable opportunity to leap forth and make a genuine contribution to improve the status quo. After all, hadn't my few isolated conversations unearthed a veritable magma of discontent roiling beneath the surface? That was doubtless a good sign. A very good sign. Now I must seize the moment and fully embrace the opportunity to become a sort of UNESCO catalyst, the outside spark that could suddenly set all that frustration alight in a brilliant burst of creative destruction. Paper-pushers of the world unite! You have nothing to lose but your staplers!

As soon as I returned to UNESCO, I made a bee-line for D'Orville's office. Perhaps it might be a good idea, I suggested warmly, to call a meeting for all key figures of the Strategic Planning Group so that I could properly introduce myself to everyone and explore future avenues of potentially innovative collaboration.

D'Orville promptly agreed to this, delighted that I was finally starting to demonstrate the possibility of becoming a good Strategic Planning team player, ordering his secretary (who had finally arrived) to set up a general staff meeting the following Monday afternoon, five days before Christmas. Ho ho ho.

Chapter 5

A Strategic Error

"*So I called this meeting,*" D'Orville pronounced grandly, "*because I think it's important that you meet Howard Burton, who's doing a consultancy with us for a few months. Now, Howard worked for many years at Research In Motion, the Canadian BlackBerry company, and I thought that he might be able to assist us in our quest to develop new innovative financing solutions, which as everybody knows has been one of my principal priorities.*"

He grabbed a thick black marker from the table in front of him, shuffled around the board room towards a waiting easel and, flipping open the oversized pad of paper, quickly wrote down "Innovative Financing" at the top of the sheet before underlining it twice. Pens on all sides of me began dutifully scratching away, recording the great man's thoughts.

"*As you all know, we've had many discussions of innovative financing in the past and are currently rigorously exploring many possible avenues: offsets, diaspora bonds, public-private partnerships, and so forth*'

He paused briefly to allow the note-takers to catch up with him.

"*But the one mechanism that I'm convinced* **doesn't** *work is a Tobin Tax,*" he frowned angrily, writing down TOBIN TAX in capital letters before emphatically putting an enormous "x" through it. All the strategic minions around me bobbed uniformly in agreement.

There must be some history there, I mused to myself as I watched D'Orville embark on yet another anti-Tobin rant. Never before have I encountered anyone—well, outside of the US, at least—who had such passionately negative feelings about a particular approach to taxation. A fervent desire to assert his own economic credentials, perhaps? Die-hard allegiance to some long-published paper? Or maybe it was more personal: an old flame summarily ditching our besotted Hans to run off with a James Tobin look-alike, or maybe even James Tobin himself. Who knows?

To make matters stranger still, my recent perusal of the "innovative fundraising literature", to use a rather comically inflated term, had revealed that the Tobin Tax was very much alive and well in all but name. UNITAID, an international drug purchasing organization founded in 2006 dedicated to combating HIV/AIDS, malaria and tuberculosis, is regarded by most observers as the poster boy of innovative financing, due to its successful imposition of an airline ticket tax (otherwise known as a "solidarity contribution") that accounts for no less than 70% of its burgeoning financial base.

This genuinely noteworthy result was directly attributable to UNITAID's then-Chairman, the former French Foreign Minister Philippe Douste-Blazy. By successfully championing this tax (together with, vitally, procuring the support of the French government in the process), Douste-Blazy, a cardiologist, became an instant international aid folk hero, universally recognized as one of the few people to have ever actually achieved anything substantial in the chattering world of innovative development financing. He was duly punished for such wonton conscientiousness by being appointed Special Advisor on Innovative Financing for Development to UN Secretary-General Ban Ki-moon, where, one imagines regretfully, he was forced to travel the globe giving inspirational PowerPoint demonstrations to whingeing administrators at countless innovation conferences.

Meanwhile, as one might imagine, UNITAID's success was followed, rather less innovatively, by a deluge of copycat suggestions for similar broad-based taxation schemes: taxes on mobile phones, taxes on internet service providers, taxes on sporting and cultural events. Not to mention, of course, James Tobin's original idea of long ago: a tax on financial transactions. None of these has come to pass—principally because of a lack of leadership at the political level: nobody of the stature, commitment and influence of Dr Douste-Blazy has risen to the fore to passionately embrace a cause and convince the powers that be to make things happen.

But a trail had nonetheless been blazed. Douste-Blazy and UNITAID have clearly shown that such broad-based revenue generating approaches not only *can* occur, but need not even be enormously widespread to have a considerable effect. For it bears mentioning that the UNITAID airport tax is only levied by ten countries, a mere four of which lie outside of sub-Saharan Africa (Chile, France, Korea and Norway).

You'd think that this point alone—tangibly demonstrating the efficacy of a tax that has struggled to gain widespread acceptance even amongst otherwise sympathetic donor nations—might go some considerable distance to quelling any virulently anti-Tobinite tendencies from those, such as D'Orville, who repeatedly cite its lack of potential universality as its overriding structural flaw. But no. As far as the Assistant Director-General for Strategic Planning was concerned, the matter was closed: it was demonstrably obvious that a Tobin Tax would simply never work and anyone who suggested it in his presence risked being subjected to the full brunt of his considerable disdain.

It was safe to say, judging by the chorus of sycophantic head-nodding and coordinated chuckles of delight that greeted every D'Orvillian pronouncement, that nobody in the Bureau of Strategic Planning was the slightest bit prepared to challenge the boss on this, or any other, significant issue. This was quite dispiriting, and went a considerable distance towards deflating my recent hopes for boldly shaking up the UNESCO establishment, or at least this particular corner of it. Whether or not a Tobin Tax was defensible or hopelessly Utopian was, of course, utterly beside the point: far more concerning to me was the fact that it was patently obvious that possible dissension from, or even casual questioning of, the official party line would be not tolerated. Rather worse still, it was hard not to conclude from my admittedly very brief impressions that nobody seemed particularly frustrated by any of this—hardly the most fertile ground to cast my revolutionary seed.

Still, we live in hope. Perhaps I would have a slightly different impression of things once I had the chance to properly introduce myself and explain my approach.

D'Orville's pedantry spluttered to a conclusion after another ten minutes or so, culminating with a plaintive request for the surrounding strategic planners to take advantage of my all-too-finite presence in their midst, a clear sign that after a few short weeks D'Orville was no less determined to see the back of me than I was of him.

"Howard will only be with us for another few months, so please make sure you take the time to talk with him while he's here. Innovative Financing is a priority for all of us and I can't do everything myself. I need your help."

On this plaintively self-pitying note, the floor was turned over to me to say a few words.

I began by respectfully pointing out that, while most appreciative of Hans' gracious introduction, it should be pointed out for completeness that I never did, in fact, work at RIM or sell BlackBerrys, or any of that. I had, however, enjoyed a rare and privileged opportunity to build a new scientific research centre from scratch, obtaining over $350 million in new monies from both private philanthropy and leveraged public sector support; and it was precisely this unique construction of an innovative private-public partnership that I imagined might be of some value to UNESCO's Bureau of Strategic Planning

In particular, I stressed, through my experiences I had necessarily developed a particularly keen awareness of the mindset of the prospective philanthropist or suitably engaged government official that lay at the heart of my fundraising philosophy.

This seemed to have grabbed their attention I gazed deliberately around the table, searching for eye contact that might flush out any sympathetic listeners, before continuing.

"In my experience," I declared, *"any successful fundraising initiative begins with a deliberate inversion of the usual perspective that I'm sure we've all seen so many times before—fundraisers approaching the issue like a construction worker trying to fill a hole: 'We need this amount of money to maintain our current budget—where can we find it? Maybe there's some government program that we can get our hands on or maybe there's some*

suitably egocentric rich guy we can find who could be enticed to give us a big whack of cash in exchange for putting his name on one of our buildings.'

*"And, as I'm sure you can appreciate, this sort of approach typically leads nowhere. Because we are coming at things from the wrong perspective: we are focusing on what **we** need rather than putting ourselves in the position of the prospective partner and asking ourselves, What do **they** want?*

*"After all, the only people who might be passionate about filling budgetary holes are those people who are already involved in the organization and are determined to ensure that it succeeds. And by definition, when one is contemplating large-scale innovative fundraising, those **aren't** the people we are targeting, because most of those people will have already contributed or somehow otherwise done their best to tangibly assist. No, we are looking here to captivate the attention of **new** people, those **outside** the current funding framework, people who have no a priori allegiance to our cause whatsoever.*

"Let's look at the situation another way," I challenged them. *"Let's suppose for the moment that each of us, everyone sitting around this table, is a corporate billionaire. Maybe we've turned a thriving family business into a global empire or maybe we've built something entirely from scratch, or perhaps we've restored a struggling corporation to its former glory. It doesn't matter. But what does matter is that, however exactly we've managed to amass our enormous wealth, we're constantly being approached by people on all sides of us anxious to have us contribute to something. Not a day goes by when we aren't assailed by someone or other asking us to participate in this or that allegedly worthy cause: make a financial contribution, sit on a fundraising committee, sign a letter, call one of our influential friends,*

and so forth. We all know how irritated most people become when their inboxes become overloaded with spam-filter-evading requests from Nigerian solicitors to provide our personal banking details. Well, we rich people get about a hundred times this sort of bombardment **each day** from legitimate charities, which has a definite tendency to deaden any philanthropic tendencies we might be inclined to have. Worse still, whenever we do succumb and donate money somewhere, the principal result is a drastic intensification in the frequency of the number of requests that we are deluged with.

"Now there is probably some fraction of us who aren't the slightest bit interested in doing anything philanthropic anyway. A few might have taken some sort of vow not to concern ourselves with this sort of thing until we reach a certain age, fearful of becoming distracted from successfully managing our companies by spreading ourselves out too thinly. Others might be opposed to the very idea of philanthropy, convinced that the act of giving money away demeans its value and discourages entrepreneurial activity, while there are those who will adamantly insist that it's the government's job to do these sorts of things rather than the private sector.

"So there will hardly be a unanimity of views, particularly when one looks at a room filled with billionaires who come from a wide variety of backgrounds. But I think it's safe to say that there will always be some people in the room who will indeed be very keen to make some sort of a significant contribution, who will want to make sure that the wealth and influence that they have likely spent the better part of a lifetime acquiring or growing, can have the greatest possible impact on the world around them. And for those of us who are keen to make a difference, the *question is clearly not: Should I do something worthwhile? But rather: How, exactly?*

"After all, we have all been around the block a few times, we boardroom billionaires. Over the years, we have witnessed, in the business world and elsewhere, scenes of waste, inefficiency and strategic failure on a huge scale. We have watched while vast numbers of people were temporarily captivated by unworkable notions, and seen many examples of good ideas, improperly implemented, go to seed. We also know from direct experience that, however important financial resources might be to getting something successful off the ground, just throwing money at a problem almost never achieves the desirable ends—and sometimes even does more harm than good in the long term.

"And the truth is that most of the people who are approaching us for funds do not make a terribly convincing case. They trot out the same old tired lines about justice and fairness and opportunity, invariably insinuating that we have a moral obligation to contribute to their cause because we are fortunate to possess so much more than most other people. Doubtless they mean well. But are they truly effective? Do they have specific goals that can be objectively measured? Do they make a real difference? Do they get us excited?

"No. Not really. Which is a problem. A big problem. Because we are more than just a bunch of rich guys: we are creators and innovators, dynamic individuals who have taken on the cut-throat competition and the naysayers and the sceptics and won. Naturally, we tend to be arrogant. Naturally, too, some of us are borderline delusional in terms of our inability to objectively assess our own qualities—many even: it comes with the territory. But that's not the point. The point is that if you want to get our attention, you have to get us excited. You're not going to get us

involved by twisting our arms, or threatening us or shaming us into it. That won't work. The only way we're going to participate is if we want to. You have to captivate us. You have to excite us.

"Because the truth is, as I said before, that many of us **do** *want to do something interesting and important in philanthropy. Many of us are excited at the prospect of making a difference and doing more in life than just increasing the market share of our companies.*

"But we are not, generally speaking, terribly impressed by the philanthropic status quo. In fact, we are so unimpressed that several of us have gone to the considerable time and trouble to start our own foundations and institutions. Of course we didn't have to do that. We could have just cut a cheque to UNICEF or UNESCO or the WHO—it certainly would have been much easier. But we didn't. Because these organizations didn't excite us in the least. They didn't convince us that our hard-earned dollars would give rise to something different, something uniquely important.

"That is why," I continued, slipping back into my own distinctly non-billionaire persona, *"when it comes to fundraising, I always talk of partners instead of donors. A donor is someone who just gives something and walks away. A partner, on the other hand, is someone who is deeply involved in something, who is excited by it, who has an emotional as well as financial stake in it. The true route to success lies in trying to generate partners, not donors."*

I paused again for dramatic effect, taking another moment to search around the room in an effort to gauge the impact of my soliloquy. The reaction seemed to be mixed: some avoided my gaze while others nodded encouragingly as they unhesitatingly jotted down my comments directly under their leader's anti-Tobin Tax rant.

D'Orville himself looked rather sceptical, although perhaps it was simply because I was talking much longer than he had envisioned.

Time to speed things up a bit.

"*Perhaps,*" I resumed, "*you think this is all obvious—looking at the world through the eyes of the prospective partner. Well, frankly, I have to agree with you—it ain't rocket science, as they say. But then we are surrounded by those who seem to have a remarkable tendency to miss the obvious. I can't tell you the number of times I was breathlessly approached by people wondering why on earth a successful tech businessman and his partners would want to fund anything so arcane as a physics institute. The real answer, of course, is that it would be quite unreasonable to expect someone like that to be involved in anything else. After all, if your entire history and worldly success revolved around understanding and exploiting the laws of nature, doesn't it stand to reason that one of the things that would excite you the most would be to push that level of understanding deeper still? Isn't that just obvious? And yet this lesson seemed largely lost on the fundraising community. The same people who ignored the tech billionaires until they began to start their own philanthropic initiatives, suddenly turned their attention on getting these new Medicis to support their projects, instead of contemplating the best fit with the next generation. And so it goes.*

"*While the constraints and motivations are quite different, as I'm sure you're all aware, related lessons apply to the public sector as well: they too need to feel enthusiastic about a project to participate, they too need to feel like partners and not donors. An effective private-public partnership has to be just that—both sides must feel justifiably integral to the experience.*

"Now some of you might feel that my unique circumstances have given me a rather distorted picture of things. After all, you might say, it's easy for you to talk so glibly about private-public partnerships. you went looking for government support holding a $120 million private-sector commitment. That's bound to get their attention.

"That's certainly true: it isn't hard to get people's attention when one comes to the table with a mittful of money. On the other hand, it's often not the sort of attention that one wants. The most common response a government official will give you if you come knocking on his door to partner on some worthy initiative with a suitcase of cash in hand is: 'That's great! Congratulations! But with all that money, what do you need **us** for? Our job is to assure proper support of the areas where there are insufficient funds, not fields where the rich and powerful are already involved.'

"So you have to make rather different arguments, of course. You call attention to how this latest investment of resources is, in fact, strongly aligned with the government's own research program and illustrate how it will clearly complement present efforts already in place. You point out that in the highly competitive world of international research, state-sponsored egalitarianism is a recipe for mediocrity; excellence only comes from a focused intensity of resources (intellectual, financial, social) that this unique opportunity provides. And you stress how a successful private-public partnership will lay the seeds for future philanthropists to come forward and invest even more resources towards the development of other laudable initiatives that will directly benefit the nation.

"Once again, these arguments are deliberately crafted to respond to the natural concerns and preoccupations of the responsible government official. Once again, the objective is *not to browbeat people into submission or shame them into supporting your cause or seduce them with the possibility of sidling up to money and power.*

"Of course the political landscape is invariably subtle and volatile and considerable astuteness is required as one attempts to navigate the corridors of power: developing champions, exploiting sensitivities, and all of that. But those are tactics. And however necessary they may be, they are still strictly secondary to our principal quest: making the case to enthusiastically bring the public sector into the private-public sector partnership. After all, that is where the champions will come from."

I paused briefly and watched D'Orville shift in his chair uneasily. Clearly, I was running well past my allotted introductory time slot.

"Anyway," I resumed hurriedly, trying to preempt him rising up to plunge into another anti-Tobinite diatribe, *"that's my basic philosophy, which I'm delighted to have the opportunity to share with you. And I'm particularly delighted to be here in the Bureau of Strategic Planning to do my best to make a contribution to UNESCO. Because, as I'm sure you can see by now, that's precisely where I think fundraising belongs. It's about money, of course, but even more significantly, it's about strategy: about **what** is to be done and **why**, about how we might expand our current thinking to involve others who can make a meaningful contribution—not only financially but also intellectually and culturally—so as to enable us to develop better programs and best ensure that they have their desired effect."*

Looking at the sea of smiling, encouraging faces around me, I began to feel quite pleased with myself, convinced that I had struck the right balance between enthusiasm and respect. I had managed to articulate a fresh and passionate perspective while avoiding any appearance of smugness or superiority, consistently invoking my audience to knowingly join me ("as I'm sure you're all aware...", "as you clearly well appreciate...") in moving forwards together to address the issues.

And by leading with my modus operandi of first focusing on the desires and orientation of potential partners, I managed to neatly sidestep the particularly thorny question of whether or not UNESCO was, in fact, presently engaged in doing anything the slightest bit worthwhile; cleverly elevating essential discussions of strategy and programmatic impact to the inspirational context of future improvements associated with fundraising opportunities. Perhaps there was hope after all? Perhaps I could actually have some non-trivial positive effect on this creaky, sclerotic behemoth?

And then Hans D'Orville got to his feet, promptly shattering my daydreams of winning the Nobel Prize for Bureaucracy Busting.

"*Very interesting,*" he said icily. "*But now that you've had a few weeks to study the matter, how exactly would you propose to go about raising the $16 billion annual supplement needed for education? That is, after all, what we're most interested in.*"

The boardroom suddenly became suffused with pointed hostility on all sides as, taking their cue from their unimpressed leader, the ranks around me swiftly closed. A potential comrade mere seconds ago, I had abruptly become the arrogant intruder, the naive know-it-all, another bitter disappointment. The great D'Orville had spoken And I had been found wanting.

"*Well,*" I began slowly, stalling for time as I desperately tried to figure out which tone to take that would be best adapted to respond to this latest attack (defiant resolve? dispassionate disdain? resolute calm?), eventually settling on matter-of-fact professionalism, "*of course, I've only started looking at this recently and hardly claim to be an expert here. But presumably one of the things you're looking for from me is my objective perspective—a fresh pair of eyes, as it were. So, since you asked, here is what I think, based on my reading of the situation.*

"I don't see that this is an area which is best suited to the idea of public-private partnership, given the amounts involved. Sixteen billion is a huge figure, and it would have to be cobbled together from a large number of distinct partnerships that would spread everyone out in all directions. Under the circumstances, then, it seems to me to make the most sense to concentrate efforts on coming up with something similar to what UNITAID did with airline tickets, which seems to have been most effective.

"But to actually achieve this, I'm convinced that, in addition to finding national champions and all that, one would first have to come up with a clear and simple message about the overriding goal so that all those being taxed would be maximally responsive to the idea—or at least minimally opposed.

"For UNITAID, of course, the message is simple: save lives. The details about how drugs are purchased, or eventually distributed, or the intricacies of how UNITAID's presence precisely affects the pharmaceutical market, are naturally irrelevant. Of course, most people in France probably have no idea that they pay a tax to UNITAID when they buy their airline tickets anyway, but even those who are aware would likely find it hard to rebel against paying a small supplement in order to save the lives of their fellow man.

"In my view, EFA needs to embrace an equally clear and compelling goal. The six separate EFA goals are too diverse and often too vague and the message should simply be reduced to that of the second Millennium Development Goal: achieve universal

*primary education: **Every child in school, everywhere**. That's a clear and compelling message that everyone can resonate with: not so different in principle from saving lives.*

"As far as what the best mechanism would be, I don't pretend to have all the answers. I know that various people have suggested mobile phones or gasoline and a whole host of goods and services"—deliberately treading as carefully as possible to avoid explicitly mentioning financial transactions or anything else that might veer ever more precipitously towards the heretically Tobinesque.

*"But personally speaking, my view is that it makes the most sense to consider an area that has a natural resonance with education, like a small supplement to the price of books, magazines, and newspapers. Or perhaps internet service providers. I'm not sure what the precise figures would be, but I'm guessing that if the EU were to propose, say, a 10 centime tax on every book, magazine and newspaper sold in Europe, with the understanding that the monies raised would go towards ensuring universal primary education, that would doubtless raise an enormous amount. And I can't imagine that it would meet with a huge amount of public resistance either: most book buyers wouldn't be terribly upset at paying an extra 10 centimes if they believed that it would enable universal primary education. Meanwhile, any awareness or marketing campaign would be based on an obvious strategy like: **'a tiny price on your book today so that everyone can read it tomorrow.'"***

I glanced over at D'Orville, who had remained motionless in his chair throughout my response, eyebrows slightly raised and arms folded across his chest like some bureaucratic Buddha. Clearly, no further engagement on his part was required after having already pronounced on my palpable lack of Strategic Planning worthiness just a few moments earlier.

"*You have a very interesting perspective, Mr Burton,*" a voice across the table from me dryly pronounced, dripping with condescension that transparently implied anything but. "*But from where I sit there is an even bigger issue, which is that UNESCO would have no idea what to **do** with an additional $16 billion even if they **could** somehow get their hands on it. I was wondering if you might comment on that.*"

Now *this* I had definitely not expected. Would I comment on the fact that UNESCO is so profoundly dysfunctional that it is manifestly incapable of doing anything productive with whatever additional resources might somehow be procured for it? How? Why? Was this a trick question, setting me up to publicly acknowledge UNESCO's wholesale ineptitude in front of the entire department? Or was it instead some sort of cutting commentary on my over-arching hubris: '*OK, Mr Smarty-Pants, who's been here all of two weeks and who seems to have all the answers, What, exactly, are we to do with all that money once you get it for us?*'

What I should have done, of course, was simply retort: "*Well, I can imagine your frustration, but as far as I understand it, you're asking me to **raise** $16 billion right now, not **spend** it.*" Or perhaps, I might have calmly sought clarification, responding with an indulgent, "*So are you asking me what I would do with UNESCO if I were suddenly Director-General?*" There was also, of course, the simple option of resolutely remaining above the fray: "*That is a very important point that needs to be fully addressed by those in charge of the organization, but in my judgement is definitely well beyond the scope of this meeting.*"

Sadly, I said none of those things. For perhaps the first time in my life, I was genuinely speechless. Here I was, doing my utmost to somehow meaningfully participate in this ludicrous game of fundraising make-believe, offering concrete advice as to how this bureaucratic mass of multinational ineptitude might conceivably meet its laughably incongruous financial targets, when along comes the Deputy Director of the Bureau of Strategic Planning—for that is who he was—to triumphantly inform me that any attempt to garner more institutional resources was a thoroughgoing waste of time anyway, since UNESCO would have no idea what to do with the money anyway.

Somehow, gauging by the sneers of delight exhibited by the others sitting around the boardroom table, this damning self-indictment was supposed to put me in my place. But it was difficult to see precisely how. As I stammered away unintelligibly, desperately trying to get my bearings in this deeply twisted world of *Howard in UNESCOland*, I kept glancing over my shoulder to see if the Mad Hatter would put in an appearance for good measure: *Tea Time!*

The meeting broke up shortly thereafter, with Hans D'Orville once again emphasizing the extremely brief nature of my UNESCO involvement and how it was correspondingly imperative for everyone concerned to take advantage of my presence in the short time remaining.

"*Too Late! Too Late!*" I muttered to myself in my best White Rabbit way, staggering out of the boardroom and back to my office to watch the prancing horses of the École Militaire down below, trying, vainly, to get a glimpse of the Red and White Knight.

Chapter 6

Burn Before Writing

Once again visions of abrupt resignations started seductively dancing through my head. My goal of unearthing secret pockets of UNESCAN dissent had come crashing to a sycophantically Kafkaesque halt within the cosy confines of Hans D'Orville's private fiefdom, and there seemed little point in continuing to pretend that I could make a meaningful contribution to a Bureau of Strategic Planning which was so conspicuously characterized by a stark absence of strategy, planning, or—rather more problematic still—basic logic.

It seemed that the only reasonable solution was to stride over into the Great Leader's office and promptly release us both from the stifling confines of this mismatch made in hell. Tomorrow, I could even spend the whole day at the Louvre and try to convince myself that the whole sorry business never happened.

And yet I hesitated.

For the truth was that, putting aside my feelings of personal frustration, it couldn't be denied that I was in the midst of a unique opportunity to peek behind the curtain and get an insider's view of this bizarre and twisted land of bureaucratic doublespeak, this hulking miasma of international inanity posing as nothing less than the flag-bearer of contemporary humanitarian values.

I do like travelling; and while delving into the bowels of UNESCO had hardly been a lifelong ambition for me, there was unquestionably something intriguing about finding oneself inside an institution that had triumphantly managed to bring hypocrisy and dysfunctionality to such Olympian levels. Excellence is excellence, after all; and all of my initial UNESCO experiences to date had amply demonstrated that I presently found myself involved in a place that could make a serious claim for being the most incompetent and irrelevant organization in the history of mankind—a deeply impressive accomplishment in its own way.

I couldn't help wondering what factors were responsible for all of that. How did it get this way? Where did it all go wrong? Was it doomed from the start? Was it remotely conceivable that it could ever, be in any way improved, even just a little bit?

After all, perhaps it wasn't actually as monolithically incompetent as I believed. Perhaps somewhere buried deep within that concrete morass of benighted sanctimony, lay isolated pockets of genuine accomplishment and impact, somehow marring the otherwise perfect record of non-achievement. Unlikely, of course. But logically possible. And if I left now, how would I ever know?

No, I should at least stick around for the length of my contract: *carpe diem* and all that. Here was my chance to find out what life inside the UN system was all about: a once in a lifetime opportunity—God willing—to objectively investigate what UNESCO is really like.

Of course, there was precious little hope that I could make Hans D'Orville happy by raising large sums of money to patch

UNESCO's justifiably listing administrative boat. But then, that was never really in the cards anyway and it's most doubtful that even D'Orville seriously thought that might happen. I was, somehow, dumped on him by the powers that be; and he responded by giving me a short-term consultancy in an effort to keep everyone happy.

Well, then, as a consultant, I would consult. Widely. I'd spend my time talking to as many people as I could, trying to get a sense of what UNESCO actually does, what it might conceivably do in the future, and what, correspondingly, it should focus its future fundraising efforts on—should those even, in fact, be objectively justified at all.

And then, like any good consultant, I'd write a report about all of that, detailing my experiences and proffering recommendations: a report that would doubtless go entirely unread, placed unceremoniously in a secure room filled to the brim with countless other carefully bound commentaries: UNESCO's own version of the final scene of *Raiders of the Lost Ark*.

But none of that mattered in the slightest. After all, there was no point in pretending that I could possibly effect any change to my surroundings one way or the other: the most that could be hoped for from the entire exercise would be a broadening of my own understanding of what goes on inside these imposing fortresses of inertia-riddled bureaucracy. I would be a modern-day Gulliver, travelling through the mystifying domain of UNESCOland, observing and recording native customs and rituals for my own edification.

And throughout it all, my investigatory musings would be anchored by one central illustrative theme: What, if anything, did UNESCO do that was actually **worth** supporting?

More precisely, and somewhat more provocatively: What would happen if the whole place suddenly disappeared? If one morning I were to amble along the Avenue de Lowendal and was abruptly confronted with the surprising, yet not altogether displeasing, prospect of discovering that UNESCO headquarters had completely burned to the ground, what would be the actual, real-world consequences for global efforts in science, education and culture? Which specific programs would need to be rebuilt? Would anybody even notice?

Yes, I enthusiastically concluded, that was the ticket: I would mentally commit UNESCO to the flames; and then, carefully and dispassionately, sift through the charred remains to see what, in fact, would need to be reconstructed. Not because I was specifically asked to do so. Nor because anybody would ever pay the slightest bit of attention to my conclusions. But simply because I was curious.

I even had a title picked out already: *Burning Down UNESCO: A Guide to Innovative Fundraising.*

Catchy, no?

Chapter 7

Historical Investigations

Before burning our present-day UNESCO to the ground, it seemed reasonable to try to develop a clearer understanding of its history and original mission in order to assess to what extent it has lived up—or down—to its founding expectations. After all, such organizations don't simply spontaneously spring forth, fully formed, from the head of some multinational Zeus. At its birth, there must have been some clearly perceived goal, a driving philosophy, a mission statement. In the white heat of its creation, we must surely find vision, a dream, hope—something.

And so we do. Of a sort. But in true UNESCO tradition, even its very origins were characteristically murky: a muddled mass of contradictory sentiments and confused ideology, a tangled amalgamation of verbose Utopianism, creaking multilateralism and resolute realpolitik.

The story begins in 1922 with the founding of the International Committee for Intellectual Cooperation (ICIC), an advisory body of the League of Nations consisting of various elite representatives of the global intellectual firmament such as Marie Curie, Albert Einstein, Thomas Mann, Henri Bergson, Béla Bartók and Paul Valéry. The idea was that, by focusing on illustrious individuals rather than political appointees to a multinational process, the ICIC would be able to elevate itself above the knee-jerk nationalistic fray and provide strong moral stances on controversial issues of the day so as to help ward off a repeat of the appallingly needless butchery of the First World War.

This both worked and didn't work. Like any group of celebrated intellectuals the ICIC was able to formulate a few solemn pronouncements on world affairs (at least when its members weren't spending their time squabbling with each other—Einstein and Bergson famously locked horns on more than one occasion, with the French philosopher laughingly determined to lecture Einstein on time, which somehow didn't manage to comprehensively destroy his intellectual reputation).

But any formal declarations from a group of celebrated intellectuals were, then as now, particularly easy to be ignored by the world powers. Suffice it to say that the International Committee for Intellectual Cooperation had no more of an effect on the irrepressible expansionist inclinations of Adolf Hitler and Benito Mussolini than the League of Nations did—which is to say, none at all.

Then came the even greater butchery of the Second World War, followed by a renewed desire to construct effective multinational mechanisms to override humanity's perverse determination to wantonly extinguish itself. During the 1945 founding conference of the United Nations, there was a recognized need to create a parallel culture and educational organization along with its primary political counterpart, encapsulated by Clement Attlee's lustrous words that were to launch the preamble of UNESCO s Constitution: *"Since wars begin in the minds of men, it is in the minds of men that the defenses of peace must be constructed"*. In 1946, the now-defunct ICIC was thus formally absorbed into a new United Nations body that would have an international mandate for culture, education and—thanks to the active lobbying of a few influential British scientists—science. The United Nations Educational, Scientific and Cultural Organization was born.

But what sort of place would it be? How would it operate? Opinions, naturally, differed. The French, who had hosted the ICIC in Paris for the past 20 years, were inclined to develop a model based on strong non-governmental representation of independent bodies. The British and the Americans, meanwhile, favoured an intergovernmental approach with control exercised by Member States. In the end, a familiar sort of multi-lateral compromise was reached: the new institution would be run according to the structure proposed by the British and the Americans, but the headquarters would stay in Paris.

And then there was the question of content. While everyone recognized that UNESCO's prime mandate was to serve as an international beacon to "construct the defenses of peace in the minds of men", as Mr Attlee so nobly put it, the devil was very much in the details as to precisely how best to achieve this. In an era when much of the world's citizenry had directly experienced torrents of propaganda coupled with the shackles of censorship, there was a natural determination to use UNESCO to create an open society founded on the principle of a correspondingly free flow of ideas.

But even this rather benign sentiment was not without complications, as the rapidly rigidifying Cold War led some to sceptically view this notion of an open society as little more than a hidden US agenda to turn UNESCO into a sort of multi-national *Voice of America* aimed at the Soviet Union and its satellite states—a not entirely unplausible prospect under the circumstances, given that, unlike the United Nations itself, the Soviet Union did not join UNESCO until after Stalin's death in 1953.

So ambiguities regarding UNESCO's precise mission abounded right from the very beginning. It was left to Julian Huxley, the eclectic English evolutionary biologist who became UNESCO's first Director-General (and brother of the famous novelist Aldous Huxley), to spell out the new organization's ideological framework in detail. This, it cannot be denied, he was particularly enthusiastic to do.

Huxley was a notoriously passionate and energetic fellow. After having made seminal contributions to a wide range of biological topics, from bird courtship rituals to the derivation of a mathematical relationship to relate the growth of different parts of an organism, he took up the post of secretary to the Zoological Society of London, where he occupied himself with, among other things, filmmaking (he directed what many consider to be the first nature documentary, *The Frivate Lives of the Gannets,* which won the 1937 Oscar for Best Short Subject), journalism, broadcasting, lecturing, and creating scholarly works on evolutionary biology, culminating in the classic 1942 text *Evolution: The Modern Synthesis.*

Rather more controversially, Huxley was also an unapologetic eugenicist, fervently believing that science should be concerned with "improving" the genetic makeup of humans. Huxley's eugenic beliefs were considerably more subtle than many of his peers (suggesting that the word "race" be replaced by "ethnic group", while also accepting a strong role for environmental factors in human development), but he naturally could not entirely escape from making spectacularly whinge-inducing remarks like: *"No one doubts the wisdom of managing the germ-plasm of agricultural stocks, so why not apply the same concept to human stocks?"*

Curiously, authoring such remarks didn't seem to disqualify Huxley as a driving force in the shaping of UNESCO, an organization dedicated to the promotion of world peace that came into being only months after the Nazis had been forced to end the most appalling eugenics experiment in human history.

Quite the contrary, in fact. Far from being sidelined, Huxley became Executive Secretary of the Preparatory Commission of UNESCO on March 1, 1946; and he and the equally eclectic British biochemist Joseph Needham, a self proclaimed Christian Marxist who went on to become the foremost Chinese science historian of his age, were hugely influential in convincing the Commission's President, Clement Attlee's Education Minister Ellen Wilkinson, that UNESCO should directly incorporate a science mandate along with its education and cultural ones—thereby "putting the 'S' in UNESCO", as it was sometimes said.

By the end of 1946, when the Preparatory Commission dissolved itself to formally establish UNESCO, Huxley had taken the reins as its first Director-General, while Needham became its Founding Head of Natural Sciences, an arrangement that lasted for only two years, as both men left the organization in 1948.

But they certainly left their mark.

Needham, spurred on by his own experiences in China during the war, consistently propounded his so-called "Periphery Principle", according to which the principal function of the new organization should be to promote scientific exchange between the "bright zones" of the developed world with the less advanced nations "on the periphery".

Something was desperately needed, he averred, to counter the established "laisser-faire" orientation of the 'bright zone" establishment who, by their arrogant determination that all academic interchange should be solely driven by perceived scientific need, were really arguing for the perpetual supremacy of their own elite institutions and corresponding old-boy networks.

Julian Huxley, meanwhile, was considerably more radical still. In a meandering 60-page essay entitled "UNESCO: Its purpose and its philosophy", Huxley begins by examining the institution's newly-minted constitution and notes that UNESCO's purpose *"is specifically laid down as that of advancing, through the educational and scientific and cultural relations of the peoples of the world, the objectives of international peace and of the common welfare of mankind, for which the United Nations Organisation was established and which its charter proclaims."*

He then claims that the constitution spells out three principal methods for achievement of these goals:

1. Using mass communication and international agreements to advance mutual knowledge and understanding of peoples
2. Educating children and spreading culture
3. Maintaining, increasing and diffusing knowledge

So far, so uncontroversial, albeit typically vague. But then things suddenly lurch towards the very peculiar.

"In order to carry out its work," Huxley informs us, *"an organisation such as UNESCO needs not only a set of general aims and objects for itself, but also a working philosophy, a working hypothesis concerning human existence and its aims and*

objects, which will dictate, or at least indicate, a definite line of approach to its problems. Without such a general outlook and line of approach, UNESCO will be in danger of undertaking piecemeal and even self-contradictory actions, and will in any case lack the guidance and inspiration which spring from a belief in a body of general principles."

In other words, Huxley assures us, simply eliminating xenophobia, providing universal access to education, and "establishing productive co-operation among the nations in all branches of intellectual activity" will never happen unless we first universally adopt a comprehensive philosophical framework to firmly embed our collected sense of morality. Otherwise, how are we to know that we are moving in the right direction?

The stage is now set for Huxley's unfettered flight into deepest Utopia. Not content with being appointed the founding Director-General of a new United Nations organization shortly after the conclusion of the worst period of widespread death and destruction in recorded history, Huxley has instead seized the opportunity to refocus attention on his all-encompassing theory of societal progress.

One can't help but sense a definite *Jerry Maguire* tinge to all of this: Huxley passionately scribbling away on his Mission Statement at 4 am, frantic at the chance to communicate his philosophical epiphany to the world. But what does he have to say? What is Julian Huxley's equivalent of *Fewer Clients, Less Money*?

Well, it's naturally a bit more complicated. Given UNESCO's deliberately-crafted role as a non-denominational global connector, it clearly has to eschew allegiance to any one theology, politico-economic doctrine or historical narrative. So, in accordance with its natural preoccupation with peace, security and human welfare, UNESCO's outlook, we are told, must *"be based on some form of humanism"*.

But what particular variety of humanism might that be? What we need here, he claims, is a world humanism, a scientific humanism. Moreover, we are informed, it *"cannot be materialistic, but must embrace the spiritual and mental as well as the material aspects of existence, and must attempt to do so on a truly monistic, unitary philosophic basis"*. Well, it's hard to argue with that—largely because it's nigh on impossible to get any clear inkling whatsoever of what Huxley is actually talking about.

The fact that a grand philosophy of humanism doesn't contain any proper definition of its key terms at the outset is decidedly troubling, but this otherwise conspicuous omission is quickly overshadowed by what follows: Huxley's bold fusing of the basic tenets of humanism with evolutionary biology. For what UNESCO needs, Huxley's triumphantly unveils, is something quite different from what has ever gone before: **evolutionary humanism**.

"Thus the general philosophy of UNESCO should, it seems, be a scientific world humanism, global in extent and evolutionary in background."

Lest anyone be tempted to conclude that Huxley's philosophical approach is only descriptive, a metaphorical interpretation of an established doctrine that simply reflects his evolutionary biological roots, the record should be set straight: evolutionary humanism is not only a radically new world philosophy—for UNESCO, it is quite simply the only game in town.

"It is essential for UNESCO to adopt an evolutionary approach. If it does not do so, its philosophy will be a false one, its humanism at best partial, at worst misleading."

The stakes are clearly high. But what the hell are we actually talking about?

Evolution, says Huxley, consists of three distinct phases: the inorganic, the biological and the human. The inorganic phase concerns itself with the expansion of the physical universe, the biological phase is limited to the speciation on earth due to natural selection, while the human phase is centred around the change in man's environment through developments in our understanding that are concretely implemented within our evolving social structures.

Evolution happens in each of these three distinct realms, but as we move from the inorganic through the organic and eventually towards the human, both the rate of change and level of complexity increases drastically.

So it is that the basic fusion processes in a star result in it having a lifetime of 10^{12} years, while complex biological systems evolve according to natural selection on timescales of 10^7 years and highly sophisticated technological revolutions in our societal fabric (internal combustion engine, atomic fission) are now occurring every 10-20 years or so. Evolution is constantly accelerating, producing ever more complex and sophisticated results.

Meanwhile, Man is the sole heir of this evolutionary process not simply because he happens to find himself at the top of the biological food chain, but because he is the only organism capable, through harnessing his intellect and placing himself in a productive societal network, to successfully engineer this last category of rapid, widespread, revolutionary change.

For Huxley, evolution is an essential piece of the puzzle because it naturally allows us to clearly define what we mean by "progress". Since progress is necessarily associated with increasing complexity, by stimulating evolution towards more sophisticated advances in the human sector, we are doing nothing less than stimulating societal progress. And this is precisely the role that Huxley imagines for UNESCO: a global catalyst for societal progress.

If the essay ended here, after eight flowery pages of highfalutin prose, there would be little to rebuke Huxley for. His hand-waving generalizations would have been gently regarded as mere inspirational scaffolding for a new multilateral organization that was determined to do its part to nudge mankind towards its hitherto grossly ignored nobler side.

But Huxley did not, unfortunately, stop there. As a "corollary" of his analysis, he claimed that the only certain means for avoiding a future war would be a single world government, a charge that was not likely to (and didn't) sit well with the Americans, among others. He droned on about how *"UNESCO must devote itself not only to raising the general welfare of the common man, but also to raising the highest level attainable to man."*

And then, like two trains careening towards each other in excruciatingly slow motion, he inevitably began to hold forth on his pet topic: eugenics. Musing on how UNESCO might be able to *"reconcile our principle of human equality with the biological fact of human inequality"*, he recognized the important work done by Officer Selection Boards during WWII, before speculating on the happy day when psycho-physical awareness could advance to the point where *"certain types of men should be debarred from holding certain types of positions"*, encouraging UNESCO to "*actively support all studies and all methods which can be used to ensure that men find the right jobs and are kept away from the wrong jobs*". While admitting that "*the preservation of human variety should be one of the two primary aims of eugenics*", he simultaneously—and decidedly oxymoronically—held that "*the other primary aim of eugenics should be the raising of the mean level of all desirable qualities.*"

As is often the case in these sorts of situations, his creepiest comments actually involve an expression of detached support for something obvious. While recognizing that "*human beings are not equal in respect of various desirable qualities*", he notes that the causes of this disparity are likely a combination of genetic and environmental factors before concluding:

"*It is therefore of the greatest importance to preserve human variety; all attempts at reducing it, whether by attempting to obtain greater 'purity' and therefore uniformity within a so-called race or a national group, or by attempting to exterminate any of the broad racial groups which give our species its major variety, are scientifically incorrect and opposed to long-run human progress.*"

Not **monstrously malevolent**, you understand, or **unimaginably inhumane**, or even simply **wrong**, but rather simply "*scientifically incorrect and opposed to long-run human progress*". One can't help be reminded of Hannah Arendt's comments on the banality of evil. And all of this in late 1946, while some of the few surviving members of the Nazi genocide were still in Displaced Persons camps across Europe.

Ah, well; as Arthur Koestler so famously said, "*You can't help people being right for the wrong reasons.*"

But Huxley was hardly done yet. All of the above was bundled into Chapter 1: A Background for UNESCO, and it is in the remaining 38 pages of the second chapter concerning UNESCO's program where he becomes positively unhinged, giving his frenetic and passionately upper-class whackiness a frighteningly free reign: equating anti-vivisectionism with crude spiritualism; arguing for the incorporation of theology, history, sociology and classical literature under the rubric of Natural Sciences; highlighting the need for increased research in parapsychology, yogic mind control, and, of course, eugenics.

It goes on and on and on. And on. Here he laments the miserable state of affairs when *"art is neglected by the dominant class or the authorities, becoming art for art's sake instead of for life's sake, so rootless that it ceases to have any social function worth mentioning"*. There he bemoans the fact that *"for the great majority of English-speaking peoples, the word 'pictures' now means the films, and the escape from reality which is the aim of most films to provide; while the real pictures, the paintings which can give a deeper and more extended insight into reality, remain largely unvisited in museums and galleries."*

He informs us of his understanding of Einstein's overthrow of the Newtonian world view; he recognizes the importance of Dalton's discoveries; he is sensitive to both Freudian and Jungian interpretations of psychology; he is moved by Giotto's frescoes in the Scrovegni Chapel at Padua; he thrills to the *"extremely explicit distillation of conscious experience"* in Bach's Mass in B minor; he appreciates the humour of Beethoven's Verlorene Groschen and the satirical contempt of Gulliver's Travels; he quotes Pope, Coleridge, William Rothenstein, Sir Stephen Tallents and John Grierson; he invokes *'the primitive but striking art of the Melanesians of the Pacific"* as well as *"the more sophisticated art of Bali"* to tangibly demonstrate his profound awareness of the subtleties of non-Western cultural expression.

It is all, in a curiously ironic sort of way, deeply impressive: the man's bombastic abilities seem to be nothing less than superhuman in extent; and it struck me that however mindlessly self-aggrandizing and comically detached from reality I had found 21st-century UNESCOvites to be, their level of delusional self-importance paled by comparison to their frothing Founding Father.

Clearly a precedent for all UNESCO-related propaganda had been set from day one: all material should be needlessly long and frighteningly unintelligible, saturated with monumentally delusional levels of UNESCO's overwhelming importance for the destiny of mankind. It was not terribly difficult to trace a direct line of descent from Huxley's pontificating prose to the turbidity of the 2010 Global Monitoring Report. History matters.

But at least it must be admitted that Huxley made the effort to lay it on the line, explicitly—if strikingly incoherently—addressing what UNESCO was for and what he thought it should accomplish through the determined application of his "evolutionary humanism".

We may disagree with the values of evolutionary humanism or remain unswayed that its implementation will actually achieve very much. We may even admit that we don't, in fact, have any genuine understanding of what it is. But it cannot be denied that Huxley's enunciation of his beliefs does provide a starting point for the very necessary discussion of why the organization should exist and what we believe it should accomplish: *Here are the goals and here is how we'd like to get there.*

Much more remains to do than this, of course: measurable targets must be created and objectively checked to see if progress is, in fact, occurring, by whatever criteria we choose to invoke—but all of this motivational preamble is, in principle at least, a good—indeed necessary—beginning.

In publicly formulating his quixotically rambling views, Huxley clearly understood this—recognizing that the act of detailing UNESCO's purpose and philosophy was even more important than the specific claim of what it should be.

And Needham, while consistently exhibiting deep scepticism of what he called Huxley's "science opium", naturally appreciated this too—hence his determination to, in turn, emphasize his own more pragmatic, operational role for UNESCO as an independent knowledge transfer mechanism from "bright zones" to "dark".

Needham's successor, the French physicist Pierre Auger, careened back towards a more idealistic perspective. In his 1952 lecture, "The Methods and Limits of Scientific Knowledge", he anticipated the cultural divide that C.P. Snow was to famously illustrate a few years later, advocating that we must *"lay the foundation for a new humanism that would be total—including science—and would take the place of classical humanism, which was also total in its time."*

René Maheu, the philosopher who served as UNESCO's sixth Director-General throughout the 60's and early 70's, and whose overriding priority was the attainment of universal literacy, admitted that operationally UNESCO's ideology was *"scientific rationalism deriving from both positivism and evolutionism"* but that it was essentially a form of secular humanism anchored by the Universal Declaration of Human Rights.

Well, that was then. But what do the leaders of today's UNESCO think?

The recently appointed head of Natural Sciences, formerly a key intellectual presence within the organization, is Gretchen Kalonji, an American engineer who has achieved considerably more renown for academic scandal than scientific leadership (suing MIT for denial of tenure due to gender discrimination, receiving a controversial, high-paying job at the University of Santa Cruz while the common-law partner of then UCSC president Denise Denton, suing the estate of Ms Denton after it was revealed that she was left out of her will following Denton's suicide). Kalonji seems to be a competent materials scientist and is a consistent advocate for the laudable goal of increasing the number of women in science. She also, I'm told, speaks the East African language Kiswahili. Given my own lack of familiarity with Kiswahili, it is, of course, conceivable that she has already gone to some lengths to enunciate her own version of an innovative, inspiring and intellectually compelling role for UNESCO's Natural Science division in that language, but I can safely assure you that no such commentary exists in English. The fall from the heady, pioneering days of internationally recognized figures such as Needham and Auger is precipitous indeed.

And what of the new Director-General herself?

Towards the end of her victorious, wafer-thin campaign over Hosni Mubarak's virulently anti-semitic protégé, Irina Bokova unveiled her own intellectual credo for UNESCO: unsurprisingly, it was to be nothing less than ... a New Humanism.

So far, so familiar. From Julian Huxley's new evolutionary humanism to Pierre Auger's new total humanism to René

Maheu's new secular humanism, we've most definitely seen this UNESCO movie before.

But perhaps Ms Bokova had something new to add. As the tenth Director-General of UNESCO and its first woman, she was, after all, ideally placed to make a new and meaningful contribution to this ongoing discussion, perfectly poised to use her leadership position to refocus global attention on what UNESCO stands for and how, precisely, it can make a difference in today's complex and rapidly changing world.

What, then, does the new DG exactly mean by new humanism, I wondered? It was time to find out.

Chapter 8

Drowning in Humanism

Following her appointment as the head of UNESCO in the fall of 2009, Irina Bokova abruptly ceased making further public pronouncements on her guiding UNESCO philosophy—fully concentrating, one imagines, on the more pragmatic task of assembling her own team of like-minded New Humanists to triumphantly move UNESCO forwards towards increased global impact.

It is only during the summer of 2010 that mention of "new humanism" once again surfaces, first in a puffily promotional "UN News Centre" interview late that July.

"You have talked about a 'New Humanism' to define what should be UNESCO in the 21st century," queries the interviewer. *"What did you mean?"*

"It's a modest idea that I had," responds our new Director-General, before describing how she visited 47 countries during her election campaign and witnessed widespread poverty, which gave her considerable food for thought.

"Why do we put more emphasis on the economy and profit, forgetting the human dimension? I think the United Nations and its development agenda which they work on, and which UNESCO contributes to, is focused on the creation of a more inclusive

society in which all humans have a chance to access knowledge. This is the idea of a new humanism, to integrate the development agenda. For example, 2010 is the year of biodiversity for the UN, with the idea that there is no life on Earth without biodiversity. Well, is it not the same thing with cultural diversity? Humans, can they live without cultural diversity? That's the new humanism.

"*Promoting gender equality and women's rights, for example, is one of the most humanistic priorities there can be for the 21st century. Or our attitude to climate change. It is also a humanist agenda that puts people at the heart of concerns.*"

One can't help but feel one's inner Texan beginning to restlessly stir during all of this, rapidly concluding that her "idea" is, in fact, considerably more modest than she publicly admits.

For the point, of course, is not to seriously question whether or not eradicating poverty or enhancing biodiversity or promoting gender equality are activities worth doing. I think it's safe to assume that the vast majority of reasonably enlightened people on Planet Earth would soundly concur with these statements, just as most of us are also strongly in favour, in principle, of peace on Earth and goodwill to all.

But the key question is, **How**, *after all is said and done, is UNESCO going to make a strong contribution towards the actual* **achievement** *of these goals?*

How will your particular approach enable it to improve upon past performance? How, in other words, will new humanism, or whatever you want to call your particular operational philosophy, actually make the slightest difference?

Because the whole reason for *having* a philosophy in the first place, it sadly needs to be emphasized, is to provide a clear, guiding framework for what one wants to achieve and why.

If one is primarily motivated to bridge the cultural divide between the arts and the sciences, then it makes sense to call for the adoption of a "scientific humanism" that explicitly represents a melding of the two traditions. If universal literacy is your preeminent goal, than you could do worse than consciously adopt a secular humanistic worldview that is anchored in the Universal Declaration of Human Rights.

But there is more to it than this. Because possessing a well-defined operational philosophy not only allows one to chose operational priorities and key objectives, it also allows for a coherent and meaningful debate on the most effective way forwards. By clearly delineating one's specific approach, one gives others the opportunity to constructively disagree without necessarily being labelled as fascists or imperialists or misogynists or whatever.

Bokova's New Humanism, meanwhile, clearly does none of this. Instead, it merely cites a laundry list of platudinously obvious desirables. By the end of 2010, she has honed this self-righteous vapidity to perfection. In a speech on occasion of the 65th anniversary of UNESCO she declares:

"My message today, in times of increasing globalization, of greater connectivity and also of rising uncertainty, with new economic, financial and social challenges, is the call for a New Humanism.

"For me, a New Humanism embodies both the culture of peace and the culture of human rights, including the freedoms of information and expression.

"New Humanism is about achieving the Millennium Development Goals—the most humanistic agenda we have ever set.

"It is about responding to climate change in ways that are sustainable in the future development of all.

"It is about protecting and promoting cultural diversity, multilingualism and its relation to biodiversity.

"It is about ensuring the rights of indigenous peoples and making the most of their expertise and wisdom.

"New Humanism points to what I see as the central question raised by globalisation today: how can we manage diversity at a time when our societies and our cities are becoming more complex and more diverse in all aspects?"

Even within this gooey mass of obscurity, concrete contradictions abound: What, in God's name, does the freedom of expression have to do with responding to climate change? What does universal primary education or reducing maternal mortality (MDG 2 and 5), say, have to do with "the central question of managing diversity"? Not much, evidently; but that doesn't matter one jot. Because it's patently obvious that Director-General Bokova and her troupe of self-congratulatory UNESCO acolytes have long moved passed the time when words meant anything at all.

Which, from Bokova's perspective, is likely very good news indeed. Because the one occasion when she actually tried to make a semantic link between her New Humanism and established historical reality—during an October 2010 trip to Milan for the award ceremony of the honorary diploma in European and International Politics at the Catholic University of the Sacred Heart—it turned out to be an intellectual disaster of epic proportions, immediately vaulting her into the pole position for the Silvio Berlusconi Prize of Scholarly Achievement.

Italy, you understand, is where the humanist movement actually began. Over the centuries, the word "humanist" has come to mean so many things to so many different people that one can certainly be forgiven for concluding that by now it has lost virtually all meaning whatsoever. Indeed, had Ms Bokova elected to remain in the land of contemporary semantic humanistic ambiguity and refrained from attempting to look scholarly, she could surely have successfully gotten away with saying just about anything. But by returning to the country of humanism's birth and specifically invoking some of its celebrated participants in the clumsiest and often completely inappropriate manner, she not only brilliantly illuminated her bountiful ignorance, she manages to come across as a ridiculous poseur to boot, doubtless not the image she was striving for.

I hardly claim scholarly expertise here myself, but that only makes the contrast between Bokova's world and reality that much starker. Take my word for it: the following five-minute summary on the origins of Renaissance humanism is readily available to any motivated soul with an internet connection or a library card. Feel free to check for yourself.

Most experts seem to agree that Petrarch got the humanist ball rolling when he conspicuously compared the literary and cultural achievements of classical Roman times to what he deemed to be "the Dark Ages" following the Empire's decline. For Petrarch, Cicero represented the acme of intellectual achievement and edifying rigour; he consistently lamented the present-day decline of both sloppy Latin and unthinking scholasticism, urging his compatriots to return to a program of rigorous scholarly pursuit akin to what Cicero delineated in his *Pro Archia Poeta*: five distinct fields of study making up the so-called *studia humanitatis*: grammar, rhetoric, poetry, history and moral philosophy. The celebrated Italian author Giovanni Boccaccio was strongly influenced by Petrarch's work for years and finally had the opportunity to meet him in 1450 when Petrarch passed through Florence. The two enjoyed a life-long friendship thereafter, corresponding regularly and meeting as often as they could. After his unsuccessful attempt at mastering Greek from the acerbic Constantinople intellectual Barlaam of Calabria, Petrarch urged Boccaccio to take up the torch, which he duly tried to do with Barlaam's student, Leonzio Pilato. Full access to the wonders of the classical world, the two realized, would naturally require Greek alongside a vastly improved knowledge of Latin.

But it took Coluccio Salutati, several decades later, to really turn humanism into an active movement. A strong admirer of both Petrarch and Boccaccio whose refined use of Latin earned him the nickname "The Ape of Cicero", Salutati was hardly a lone, wandering intellectual like Petrarch or Boccaccio, but rather the long-serving Chancellor of Florence who used his eloquence and literary skill to emphatically contrast Florentine values of liberty and republican freedom with the tyrannical regime that consistently threatened it with extinction: the autocratic Gian Galeazzo Visconti of Milan.

When he wasn't trying to rally Florentines to the defensive of their city-state, Salutati took advantage of his power and prestige to directly encourage talented young scholars to come to Florence and actively engage in the great rediscovering of the classical past that Petrarch had initiated, the single-minded pursuit of Cicero's *studia humanitatis*. These brilliant, often rebellious, thinkers thus became known as the *umanisti*—the humanists, and their proud rallying cry was *ad fontes*—to the source.

In their view the way forwards, as it were, was first backwards—towards a rigorous appreciation of the clear prose and penetrating insights of the classical era, so superior to the lamentably poor educational standards of the present day.

The final feather in Salutati's humanistic cap came when he recruited the renowned Greek scholar Manuel Chrysoloras from Constantinople to his thriving humanistic circle. By the end of the 14th century, Florence was the new intellectual centre of the world.

Leonardo Bruni, Poggio Bracciolini, Niccolò Niccoli, Pier Paulo Vergerio and Roberto Rossi all cut their humanist teeth in Salutati's Florence, while other pockets of humanism began breaking out across Italy. In 1440, the brilliant and formidably irascible Lorenzo Valla demonstrated a surprising application of all of this scholarly rigour by proving that the *Donation of Constantine*—a document where Constantine apparently delegates control of the Western part of the Roman Empire to the Catholic Church—was actually a forgery. Valla established this on philological grounds, cleverly demonstrating that the Latin used in the document actually belonged to the 8th century, rather than Constantine's own 4th. Despite this achievement, and many other provocative acts that inevitably raised the ire of many members of the Catholic establishment (at one point, he was forced to cut short a visit to Rome in disguise for fear of his life), Valla somehow managed to end up as the ultimate curia insider, the apostolic secretary to the humanist pope Nicholas V, in what some reasonably described as *"the triumph of humanism over orthodoxy and tradition"*.

Valla's philological success in turn inspired Erasmus, almost seventy years later, to revise the Latin translations of the Bible (the Vulgate) by explicitly comparing and contrasting it with their Greek originals. While the fruits of Erasmus' ad fontes research were originally welcomed by the Catholic Church, there is little doubt that his new biblical translations, coupled with an analytic focus on the source text, had a strong influence on the coming Reformation.

So much for a brief history of the influential umanisti movement of Renaissance scholarship. Enter, now, Irina Bokova at the Catholic University of the Sacred Heart in Milan, some five hundred years after Erasmus:

"I am very honoured to receive this award and to be invited to discuss the topic of the new humanism. What better place to do so than Italy, and especially Milan, a cradle of European humanism and hotbed of the Renaissance?

"This revolution is at the very heart of global civilization. There are a thousand reasons we should look to it for inspiration. Among all the Renaissance cities of Italy—Florence, Rome, Venice, Assisi, Mantua. Padua—Milan offers a unique vision and highly original understanding of the humanist message. It could even be called a 'Milanese miracle'.

"Petrarch lived here for eight years; Leonardo da Vinci spent the most productive years of his life here (1489-1499). Others such as Bramante, Lorenzo Valla, Francesco Filelfo, and of course Poggio Bracciolini, made Milan the capital of a vigorous, acerbic brand of humanism, very different to that of Venice or Florence."

Well, you might be tempted to think, this is very inspiring stuff. The new Director-General travels to the birthplace of humanism to accept a prize in European and International Politics, and seizes the occasion to highlight the obvious continuity and ongoing resonance between her new guiding philosophy for UNESCO and its original development in the history of ideas, right in the very home of its proud intellectual forebears. After all, there are, as our Director-General pointedly declared, a thousand reasons why we might wish to look to the Renaissance for inspiration.

But only if we have any reasonably clear idea of what we are actually talking about. Because one of the striking characteristics of history, you see, is that it is loaded with facts. And a quick examination of the situation reveals that the new Director-General has violently and publicly propelled herself into a largely fact-free zone.

So, let's examine the record, shall we?

It cannot be denied that both Donato Bramante and Leonardo da Vinci spent important years of their careers in Milan. Bramante was a highly-regarded architect and painter who worked there from 1474 until 1499. Da Vinci, meanwhile, arrived in Milan from Florence in 1482 as an emissary of Lorenzo de Medici, and seized the opportunity to obtain patronage from Milan's Duke Ludovico Sforza, famously penning a self-promotional letter delineating ten separate civil and military engineering feats he was capable of, before adding—as an afterthought that surely must go down as one of the greatest understatements in the history of mankind—that he was also quite competent at sculpture and painting.

So it is certainly true that Bramante and da Vinci made extremely significant cultural contributions to 15th-century Milanese life. But however impressive their particular artistic, cultural, architectural and engineering achievements were, neither were scholars of the Renaissance humanist school which is the subject of our discussion, given that they didn't spend their time poring over classical texts or expressing themselves in Ciceronian Latin or any of that other business that qualified one as a *umanista*.

On the other hand, Petrarch, as we've seen, clearly *was* a humanist—indeed, in the minds of many, he was its founder. A notoriously well-travelled fellow, Petrarch wound up spending the majority of his fifties there. He was recruited in 1353 by the then-ruler of Milan, archbishop Giovanni Visconti, and when Visconti died the following year, Petrarch remained in the employ of the court of Milan, acting as a diplomat and emissary to the new joint rulers, the archbishop's nephews Gallazeo and Bernabò Visconti, who had an unparalleled reputation for cruelty and barbarism even by the frightfully depraved standards of the day. In fact, there had been a third brother—Matteo—reputed to be even worse than the other two, but his brothers were said to have poisoned him shortly after they came to power. Gallazeo was most famous for inventing the quaresima, a particularly sadistic form of torture, while Bernabò famously became an exemplar of the horrors of tyranny in Chaucer's *A Monk's Tale*. Petrarch, meanwhile, did not seem to have been terribly bothered by all of this and remained as their diplomatic representative until he left for Padua in 1362.

Then there was Francesco Filelfo. A highly accomplished Latinist, Filelfo moved from Venice to Constantinople in 1420 to master Greek as well. Seven years later he left Constantinople and stayed briefly in Venice and Bologna before settling in Florence, where he quickly garnered a reputation for incomparable brilliance and perhaps even greater arrogance.

During Cosimo de Medici's brief Florentine exile, Filelfo loudly called for the death penalty to be imposed upon him; and when Cosimo returned to Florence the following year, Filelfo unsurprisingly found himself withdrawing to Sienna, where he stayed for four years, before washing up at the Milanese court in 1440, still ruled by the Visconti descendants of the enchanting duo that had welcomed Petrarch almost a century earlier.

So score two points for Bokova's historicity: Petrarch and Filelfo *were* indeed two bona fide humanists who did actually live a portion of their life in Milan, albeit not in what most scholars consider to be their most productive years and—rather more significantly—about a century or so apart from each other. Hardly evidence for the existence of a wondrously unique brand of Milanese humanism that our Director-General maintains, it must be admitted, but at least some sort of correlation with the historical record.

Which is more than can be said for the rest.

We've already met Lorenzo Valla, the irascible philologist who, among other things, founded the discipline of philology and demonstrated that the *Donation of Constantine* was a forgery. Born in Rome, Valla left the eternal city in his early twenties, spending the next five years travelling through northern Italy, including a stint as a professor of rhetoric at the University of Pavia (before being forced to leave after having antagonized many members of the law faculty). He did, it seems, at least live for a while in Milan during this time—as well as Genoa and other locales—before moving on to the court of Alfonso of Aragon's court in Naples where he spent 13 years and produced his most celebrated works, including the aforementioned proof of the inauthenticity of the *Donation of Constantine*. After that, intriguingly, he managed to procure a position as apostolic secretary to Nicholas V in Rome.

Meanwhile the unquestionably acerbic Poggio Bracciolini, born a generation earlier than both Filelfo and Valla, was a member of Salutati's shining Florentine circle of brilliant young *umanisti* before leaving the Tuscan capital to spend the vast majority of his professional career in posts at the papal curia in Rome. But never (of course!) in Milan.

A notoriously polemic fellow, Bracciolini carried on a celebrated excoriating correspondence with both Valla and Filelfo that played no small part in establishing the stereotype of the prickly academic. Had they all somehow ended up in the same city (which, in all likelihood clearly would have been Florence), not only would scholars have been denied the vicarious delights of this three-way epistolary slug-fest, it is quite possible that they would have all ended up killing each other. But they never did.

Thus of the six people that UNESCO's Director-General specifically invoked to highlight the peculiarities of "*the unique vision and highly original understanding of the humanist message that made Milan the capital of a vigorous, acerbic brand of humanism very different to that of Venice or Florence*", two were important humanists living in vastly different times whose turbulent career paths *did* take them to Milan for at least a while, two spent undeniably productive years in Milan but were clearly *not* humanists in any established understanding of the term, and two had, to all intents and purposes, essentially nothing to do with Milan whatsoever.

One is inevitably left to conclude that, from a humanist perspective, the only thing miraculous about Milan is that a Director-General of UNESCO had the good fortune to somehow get away with spouting forth such laughably spurious declarations at a Milanese institute of higher learning in front of what one can only conclude was an equally obtuse audience.

The parade of ignorance continues. Still reeling from our encounter with the "Milanese miracle", Bokova invokes the fifteenth-century philosopher Pico della Mirandola's *Oration on the Dignity of Man* as an encapsulation of the humanist message that "*has many ramifications*". While universally regarded as a particularly brilliant and original thinker, Pico's neo-Platonistic orientation and broad-based eclecticism actually put him rather to the margins of mainstream humanism. And just to emphasize our theme of the moment, he spent his short life in Mirandola, Bologna, Ferrara, Padua, Paris and Florence. But not Milan. Can a South Dakotan Renaissance Humanist Miracle be far behind, one wonders?

And from there the Director-General makes the inevitable segue to the land of all-encompassing banality that is her "New Humanism", where poverty is bad, biodiversity is good and *"being a humanist entails building bridges and strengthening the human community to take up our challenges together"*, before concluding with yet another laughingly inappropriate bastardization of the past by assuring us that *"humanists promoted the use of 'vulgar' languages to counter the uniform use of Latin"*, thereby, one imagines, demonstrating a solid historical precedent of UNESCO's current determination to *"protect cultural diversity from uniformity."*

Except—as you're probably guessing at this point—this is, in fact, the complete opposite of what actually happened. In their deliberate *ad fontes* way, the Renaissance humanists were working very hard to **promote** the use of Latin rather than counter it. And, to add insult to injury on the topic of "vulgar" languages, the "Vulgate", the version of the Bible famously translated from Greek to Latin by St Jerome and reworked, as mentioned earlier, by Erasmus, was, (*of course!*) in Latin.

It is all enough to make you suspect that some historically knowledgeable speechwriter at UNESCO was having a bit of a laugh—until it rapidly dawns upon you that UNESCO is clearly bereft of anyone actually knowledgeable about anything.

Now, I appreciate that you might think that I have become rather carried away by all of this. So UNESCO's Director-General is congenitally disposed to wallow in broadly vapid generalizations. So she gave a laughingly inaccurate speech while trying to force a link between the organization's present "operational philosophy" and a celebrated intellectual movement of the Renaissance. So what? Does that really matter?

Well, perhaps not, given that UNESCO itself doesn't seem to matter terribly much. Indeed, by all accounts we have now reached the point where one expects nothing more from a UNESCO Director-General than the mouthing of platitudes and the consistent reinforcement of the patently obvious, tasks that the blandly unimpressive Ms Bokova is clearly eminently qualified for.

But if we are to think even for a moment about changing the status quo, if we are to seriously contemplate a scenario where UNESCO might one day do something significantly more than simply preach to its dwindling little choir of "progressive" philistines, it's going to take a significant dose of real leadership to get there. It's going to take somebody genuinely thoughtful, genuinely knowledgeable, genuinely engaging, genuinely honest, genuinely focused, genuinely realistic, genuinely capable. And that person is clearly *not* Irina Bokova, however much her star might shine when contrasted with her dictator-sponsored, anti-semitic competition.

It is hard enough for a large, unwieldy, clunkily multinational, hideously hidebound place like UNESCO to re-invent itself as an effective force for good, loaded down as it is with a litany of entrenched interests and deeply congested by its own stifling inertia. But without bold, inspiring leadership, it is downright impossible.

Of course, it could well be that we are now well past the point of no return and that the present organizational structure is de facto incapable of ever producing a strong, passionate, educated, sophisticated leader with the courage of her convictions and the dynamism to ensure genuine, meaningful impact for the institution. In which case, we are left with little option other than to consign the entire place to the flames and start over.

Think of it as incendiary humanism.

Chapter 9

Lighting the Match

It's time for a brief recap. In the great humanistic *ad fontes* tradition we have gone back to the source and spent considerable effort examining UNESCO's original purpose and ambitions before vainly seeking a clear sense of its current leadership philosophy.

But there's also the question, What is the organization actually doing right now? What is happening on the ground? Perhaps it has managed to somehow muddle through to achieving significant world-wide impact without any strong leadership or guiding philosophy. It's hardly likely, of course. But it's at least logically possible.

Remember our guiding query: If UNESCO were to burn down tomorrow, which globally pioneering programs would be lost? What would any reasonably objective observer necessarily insist must be reconstructed from the ashes? In other words, what precisely is the unique, salient contribution that UNESCO makes to world-wide efforts in science, education and culture?

It's important to be coolly precise here. Our goal is not to attack any specific people or avenues of research, but merely to engage in some calm, detached analysis of UNESCO's overall global impact. Moreover, it is likely the case that, as my institutionally-bipolar friend Jacques was at pains to point out to me some time ago, there are many UNESCOvites "in the field" who are highly skilled and impressively dedicated, fully engaged to doing their very best, often in very difficult circumstances, in order to make a real difference to the lives of their fellow human beings.

That UNESCO and its vast numbers of affiliated bodies contain many such well-meaning and at least occasionally effective people is not the point. Nor is the related claim that, for all its faults, the world would be far better off with our present-day UNESCO than without it (an as-yet untested hypothesis, but one that is probably worth taking on faith for the time being, despite the Director-General's best efforts to simultaneously impugn the causes of critical inquiry and objective reality that should properly lie at the heart of any institution centred on the causes of science and education).

No, the point is not that UNESCO does *some* good. We'll cavalierly take that as a given (at least for the time being). But we are spurred on, let us recall, by the salutary realism of a fundraising perspective.

Let's return once more to my boardroom musings of Chapter 5, but now in a decidedly more realistic vein than my futilely rhetorical attempt to seek out allies among UNESCO's Strategic Planning Team. That is, we imagine that we are approaching the likes of a Bill Gates, George Soros or Jeff Skoll with the specific aim of soliciting their partnership on select UNESCO programs. Now, we know from the start (great news!) that *all* of these fellows are philanthropically-inclined billionaires who have consistently demonstrated a strong determination to personally invest in a wide spectrum of activities that fall directly under the purview of UNESCO's core mission.

What shall we tell them? How shall we make the case?

As mentioned above, it is hardly enough to declare that hard-working, capable people exist within the UNESCO envelope or that, despite everything, UNESCO makes a net positive contribution towards humanity. That is a necessary, but far from sufficient condition; and they would have likely politely assumed as much.

Nor would an effective approach be to emphasize that the global breadth of UNESCO programs is a proud reflection of the collective will of its 196 Member States. The dead weight of bureaucratic multilateralism, with its accompanying nationalistic pork-barrelling that consistently distributes resources in a depressingly distorting way, will likely only make any successful entrepreneur run screaming to the hills.

And it's certainly not a productive use of our time to approach them beseechingly, cap in hand, wailing that recent budgetary cutbacks have brought our organization to the veritable breaking point—a mysteriously popular technique amongst the NGO fundraising community despite its unsurprisingly low success rate. *"Why are you facing budget pressures in the first place?"* will likely be their eminently reasonable reply. *"And why on earth should we private citizens step in to fill a funding gap created by the diminished contributions of national governments?"*

No, none of these techniques is going to work terribly well.

Instead, we must lead proudly from our strengths: clearly demonstrating the singular effectiveness of our unique programs in obtaining groundbreaking results that specifically resonate with our shared priorities, tangibly illustrating how a meaningful partnership with them and their related organizations might concretely advance both of our goals. In other words, a partnership with UNESCO should have nothing to do with guilt, or the intrinsic values of multi-nationalism or whether or not our philanthropists in question might have one or two extra billions lying around (they do not, I can assure you—independent of however many billions they might possess). We must convince them that a partnership with UNESCO is simply the best way to advance our (shared) interests. Because if they can't be convinced of that, there's no way they will do it. And, quite frankly, there's no way that they should.

"We do something that we're both critically passionate about in a way that nobody else can," would go the unvarnished argument. *"Let's work together to make sure we can do it even better and have even greater impact."*

Of course, this approach might well still prove to be unsuccessful for a whole host of reasons: our respective goals might not, in fact, be quite as aligned as we believe, or perhaps our philanthropist has already made her financial commitments for the year and has turned her attention to other issues, or maybe there will always remain a lingering scepticism at the prospect of partnering with any large, multinational institution. There are never any guarantees.

Except one: that if we can't manage to clearly enunciate what makes us uniquely effective and singularly important, there is virtually no way that anyone worth associating with would be the slightest bit interested in joining forces with us.

It always starts with that.

So back we march, match firmly in hand, to the metaphorical UNESCO bonfire, anxious to discover the institution's unique strengths, what the world should indignantly demand be rebuilt from the ashes once we are finished our heinous deed of intergovernmental arson.

Let's start with a simple overview of the facts.

The 2012 version of UNESCO consists of 196 Member States and 8 Associate Members. Aside from the headquarter buildings in Paris, there are 53 separate UNESCO field offices throughout the world (15 in Africa, 8 in the Arab States, 14 in Asia and the Pacific, 5 in Europe and North America and 11 in Latin America and the Caribbean) and 11 distinct UNESCO Centres and Institutes. UNESCO maintains links with 322 NGOs, of which 22 are considered "formal associates" and have offices within UNESCO headquarters in Paris.

The organization has five principal program areas: Education, Natural Sciences, Social and Human Sciences, Culture and Communication and Information.

All of its activities are financed by a combination of direct contributions from Member States (regular budget) and individual, ad-hoc agreements with Member States/others (extra-budget) for specific, targeted programs or institutions, virtually inevitably located in that particular Member State.

UNESCO presents its budgets in two year intervals. The Total Operating Budget for 2010-2011 was $653,000,000 for the regular budget and $462,471,400 in extra-budgetary programs, resulting in a combined biennium budget of slightly over $1.1 billion or a net annual budget (regular and extra-budgetary) of slightly more than $550,000,000.

In terms of its five programmatic areas, the 2010-2011 biennium operating budget is broken down as follows:

$180,554,000 for Education ($118,535,700 budgetary, $62,0008,300 extra-budgetary); $244,196,100 for Natural Sciences ($59,074,000 budgetary, $185,122,100 extra-budgetary); $56,678,800 for the Social Sciences ($29,654,100 budgetary, $27,024,700 extra-budgetary); $125,126,900 for Culture ($53,749,700 budgetary, $71,376,700 extra-budgetary) and $116,571,700 for Communication and Information ($33,158,000 budgetary, $83,323,700 extra-budgetary).

$303,300,100 of the biannual regular budget of $653,000,000, or 46.4%, is directly allocated towards UNESCO programs (all five sectoral program budgets along with the UNESCO Institute for Statistics), leaving a whopping 54.6% for administration.

This imbalance between programmatic expenses and administrative overhead is more than alarming, it is simply unconscionable, and would naturally need to be addressed and somehow justified before any specific programmatic fundraising initiative. But we are getting ahead of ourselves: right now, the goal is to simply examine each of the five programmatic areas and determine what amongst them is providing an essential, uniquely important service.

Let's take them one by one.

Communication and Information:

The notion that UNESCO should have a thriving global communications agenda is both an old idea and an entirely reasonable one. Julian Huxley, in his meanderingly verbose 1946 treatise that we discussed earlier, "UNESCO: Its Purpose and Its Philosophy", notes that in the first Article of its Constitution, UNESCO is expressly instructed to pursue its aims and objects by means of the media of mass communication and enthuses about the historic opportunity to affect hearts and minds through the "revolutionary" potential of these new technologies.

Indeed, in 1946, with the devastating effect of Nazi propaganda still very much in the forefront of the public consciousness, it was clear that a preeminent method for UNESCO to "construct the defenses of peace in the minds of men" was to seize the airwaves and boldly disseminate the fruits of its education, science and cultural initiatives. While the Soviets and the Americans were steadily moving towards the establishment of a globally bi-polar propaganda war, it seemed eminently reasonable that UNESCO might play a vital role as an objective portal of substantive content for science, education and culture.

Somehow along the way, however, any focus on content was lost and UNESCO came to be preoccupied with the *theory* of communication rather than the substance.

In the 1970s, hard on the heels of the United Nations General Assembly's commitment to a *New International Economic Order* to allow the developing world a greater impact in the global economy, UNESCO focused on creating a corresponding *New World Information and Communication Order*, appointing a commission, chaired by the Irish political activist Sean MacBride, to investigate ways of making the global media scene more equitable and rid it of its "Western bias and obvious corporate interests". This the so-called MacBride commission promptly did, producing the 1980 report "Many Voices, One World" that called for a democratization of communication technology and the strengthening of national media to avoid the dependence on (largely private), external sources.

The United States, vehemently opposed to the MacBride commission's anti-private-sector bias and fearful that the report was a dangerous step towards the creation of a global multinational media body that would control information and inevitably invoke censorship, resigned from UNESCO in protest in 1984, citing freedom of the press (There was also a parallel issue over Israel that is best to ignore at this point). The UK followed in 1985 for ostensibly the same reasons, while Singapore, never one to be terribly concerned with freedom of the press (or Israel, for that matter), but clearly anxious to seize the opportunity to save some money by leaving a sinking ship of dubious value, also pulled out in 1985 due to rising membership fees. The UK, US and Singapore rejoined UNESCO in 1997, 2003 and 2007 respectively.

UNESCO backed away from the MacBride report on the heels of the controversy; and the report itself was out of print for some time.

Today, as *CNN* increasingly depends on civilian cell phone videos for their frontline news and anyone with a blog and an opinion is a foreign correspondent, this entire debate has taken on a decidedly irrelevant hue. Indeed, the primary concern now seems to be whether or not autocratic regimes will support, or at least not destroy, the modern communications infrastructure that naturally allows the democratization of information to occur. Given that some of these countries are Member States of UNESCO, one could well imagine that some tensions might inevitably result here.

But in February 2011, when I met briefly with Janis Karklins, the Latvian Assistant Director-General for Communication and Information, he vehemently denied that this was an issue at all. Moammar Ghadafi, he told me proudly, was a keen supporter of both his UNESCO programs and a modern, well-developed internet infrastructure for Libya.

One month after our discussion, as if on cue, Ghadafi shut down Libya's internet as he vainly tried to crush the insurrection that eventually managed to overthrow his bloody regime. Yet one more example of UNESCO proudly cultivating the peace.

Meanwhile, if you look at the domestic communication agenda, as it were: if you take a moment to peruse UNESCO's own website, you will suddenly find yourself drowning in a sea of unwavering self-indulgent triumphalism: UNESCO has done this! UNESCO has done that! The non-expert reader, accidentally stumbling upon UNESCO's website for the first time, might be forgiven to conclude that the place actually matters.

Credit, of a sort, for this unrelenting stream of bombast should properly go to Eric Falt, head (Assistant Secretary-General) of UNESCO's own Comunications Department.

Mr Falt is one of the few obviously non-inert members of UNESCO's Senior Management Team, and a clear expert on branding, impact, strategic communication plans and all that sort of thing. But that is not, of course, what UNESCO actually needs. What is clearly missing from its own communication agenda is meaningful *content*—a reason for those not already firmly in the UNESCO fold to go to the website in the first place. And to return.

The whole situation, I explained to Falt during our one encounter, reminds me of that scene in *Annie Hall* when Woody Allen's character, awkwardly jerking his car in random bumper-car-like directions as he leaves his lunch date with *Annie*, is bemoaning the California mentality towards culture.

"What's with all these awards? They're always giving out awards. Best fascist dictator: Adolf Hitler."

Gauging from his reaction, I have the feeling that Eric Falt is not a huge Woody Allen fan.

Meanwhile, UNESCO's Communication and Information section truncles along obliviously, convening meetings on "free and open sources software", "internet literacy", "open access to scientific information" and the like—giving out buckets of awards while contributing nothing significant to the causes of either science and education content or substantial intergovernmental policy on information and communication.

In short, there is nothing in particular that needs to be reconstructed from the Communication and Information ashes. One down.

Culture:

UNESCO's Culture sector is where its flagship program, World Heritage Sites (WHS), lives. Frankly, I've never been a huge fan of WHS. Aside from the fact that there seems to be, in principle, a potentially infinite number of such places—which would, in the fullness of time, presumably diminish the cachet of being selected in the first place—it's never been clear to me what the point of having "World Heritage" status is anyway, or what, in fact, it actually means.

But there's no denying the fact that this is the singularly most successful part of UNESCO in terms of popular recognition and resonance with "the man on the street". For most people, the World Heritage Sites Program simply *is* UNESCO. They might not feel terribly passionate about it in itself, but at least they've heard about it. Which, from a UNESCO perspective, is saying a great deal indeed, highlighting an obvious opportunity to leverage the WHS into something genuinely impactful.

Unfortunately, the trend is hardly in that direction. Instead, UNESCO's cultural boffins have opted to build upon the success of the World Heritage Sites by expanding into the nebulous terrain that is "Intangible World Heritage", a program which deftly allows them to simultaneously profit from two of UNESCO's strongest attributes: making meaninglessly frantic pronouncements about the most obscure things imaginable and awarding prizes.

In 2011, for example, we are informed that, among others, Brazil's Yaokwa, the Enawene Nawe people's ritual for the maintenance of social and cosmic order, and Mali's rite of wisdom of the secret society of the Kôrêdugaw are firmly placed on the List of Intangible Cultural Heritage in Need of Urgent Safeguarding.

"*Well*," one imagines a bemused Athenian muttering to herself before heading off to join a demonstration against the most recently imposed government austerity program, "*good luck with all of that.*"

So the World Heritage Sites legacy has been somewhat less than inspiring. But the program itself—love it or hate it—has unquestionably had its effect on the general public and continues to do so.

The Culture sector is also the UNESCO home of "major themes" such as "creativity", "dialogue" and "normative action", along with enough incoherently prolix and gut-wrenchingly sanctimonious prose to make the most liberal, peace-loving individual desperately seek out membership in the National Rifle Association. Even if they live well outside the borders of the United States.

So here's the conclusion: in the aftermath of a sudden UNESCO conflagration, the sudden lack of World Heritage Sites would most definitely be remarked upon. But nothing else from the Culture sector would be.

Social and Human Sciences (SHS):

The self-proclaimed mission of the Social and Human Sciences sector is to *"advance knowledge, standards and intellectual cooperation in order to facilitate social transformations conducive to the universal values of justice, freedom and human dignity."*

Like so much that emanates from UNESCO's motherhood and apple pie communications agenda, it's awfully hard to find anyone who might disagree with this in principle. But I also don't know anyone who can point to any SHS program that actually produces the slightest amount of global impact.

On top of this generally benign ineffectiveness, SHS is also home to some genuine weirdness, even by UNESCO's dizzying standards. There is a program here dedicated to "Anti-Doping in Sport". Where the hell did *this* come from you might ask, and how does this have to do with UNESCO's mandate, defined as broadly as possible (other than "anything")?

There is also a program in "Philosophy"—doubtless to compensate for the glaring lacunae of the field of philosophy throughout the world's universities. And then there is my personal favourite: the theme of "Human Rights-Based Approach to Programming" where the SHS is encouraged to turn its finely honed analytical skills reflexively back towards the United Nations System as a whole to ensure that *"the Universal Declaration of Human Rights and other international human rights instruments guide all development cooperation and programing in all sectors and in all phases of the programming process."*

In the first place, I can't help wondering, yet again, why these people seem so curiously compelled to make unnecessary addenda, like explicitly bringing in the prospect of "other international human rights instruments"? Isn't the whole **point** of having a **Universal** Declaration of Human Rights that everyone signs on to, **precisely** that of avoiding to worry about the muddling ambiguity of "other international human rights instruments" that could justify potential human-rights abusers?

And secondly, isn't it awfully curious that the United Nations is so nervous about whether or not its manifold programs are in accordance with a document as obviously basic as the Universal Declaration of Human Rights that it must establish a working theme in one of its own bodies to check up on it itself? As a general rule, I'm all in favour of internal review, but this one strikes me as perhaps not the optimal use of resources: after all, if we can't safely assume that the United Nations is not, through its programs, itself contributing to the violation of human rights, our whole multinational infrastructure is on pretty shaky ground in terms of overall credibility, I should think.

If the SHS were to burn down, I must confess that I would miss its delightfully mangled use of quirky intellectual gymnastics to justify its own existence, along with Pilar Alvarez-Laso, its charmingly congenial and unpretentious Assistant Director-General. But those are personal sentiments The world at large, one could safely conclude, would miss nothing whatsoever.

Natural Sciences:

UNESCO's Natural Science sector tries to cover a wide spectrum of areas, from science policy (science governance, science legislation, university-industry partnerships) to various specific science and engineering programs (science and technology education; renewal and alternative energies; anodyne promotion of International Year of X—Chemistry, Astronomy, and the like).

For the most part, it has limited to no real effect, given that it typically has no scientific or research capacity of its own to speak of and is reduced to serving the role of a "connector" between the scientific community and government, NGOs, industry groups and so forth.

Back in 1946 when UNESCO was founded, when for political, logistical or financial reasons vast numbers of working scientists had no easy mechanism for an open exchange of ideas (recall Joseph Needham's notions of scientific "bright zones" and "dark zones"), it was hardly unreasonable to envision a clear role for an organization like UNESCO to serve as an honest broker and a conduit of scientific interchange. But that time has long passed.

Today, an institution like MIT alone can do (and doubtless does) more to promote "capacity building" throughout the developing world than all of UNESCO simply by putting some of its courses online.

And yet, there are two clear aspects of UNESCO's Natural Science sector that would, indeed, be missed after our prospective inferno, namely the IHP and the IOC.

The International Hydrological Programme (IHP) is a coordinated global effort of research, assessment, education and policy advice in water management, complete with a concrete educational mechanism, the UNESCO-IHE Institute for Water Education in Delft, The Netherlands.

In addition to its overwhelming relevance for global health and welfare, water management is clearly an issue that inherently requires multi-national cooperation, given the inter-connectedness of global flow systems and water's natural tendency to be indifferent to national boundaries. Proper long-term administration of this vital resource will clearly necessitate the involvement of an intergovernmental institution; and if UNESCO were to suddenly disappear, a similar sort of structure for international water resource management would clearly have to be erected.

A similar sort of argument applies to the Intergovernmental Oceanographic Commission (IOC). The world's oceans make up some 71% of the Earth's surface and constitute perhaps the best definition of a "global commons" that one can imagine. And yet, there is no independent UN Agency dedicated to its stewardship.

The IOC, established in 1960, is the one UN body responsible for ocean science, ocean observatories, ocean data and information exchange and ocean services such as tsunami warning systems. As their website declares, "In addition, the Commission strives to further develop ocean governance, which necessitates strengthening the institutional capacity of Member States in marine scientific research and ocean management."

Doubtless the IOC would have a far easier time of "developing ocean governance" if it could somehow find itself removed from UNESCO's stifling yoke. From all accounts, the IOC doesn't benefit in any significant way from its close association with UNESCO, and there seems to be no logical a priori reason why it shouldn't exist as a grown-up, fully-independent UN Agency with its own binding Charter for its Member States, rather than reduced to the role of vaguely pathetic quasi-autonomous UNESCO agency, a perennially underfunded supplicant to the whims of its bureaucratic paymasters who seem far more preoccupied with political posturing than actually seizing one of the very few arrows in their quiver to develop positive global impact.

There is thus a serious argument that the putative burning down of UNESCO would actually be in the long-term strategic interests of the IOC, finally giving it the impetus to rebuild itself as the fully independent UN Agency that it likely should have been in the first place. But it would certainly need to be rebuilt, one way or the other.

There is a third aspect of UNESCO's Natural Science sector which would normally merit our consideration here: The International Centre for Theoretical Physics (ICTP) in Trieste. The brainchild of the Nobel Prize-winning theorist Abdus Salam, who managed to establish the Institute in 1964 through a tri-partite arrangement with the International Atomic Energy Agency, UNESCO and the Italian government, the ICTP has carved out an impressive niche for itself as an active international research centre that is strongly oriented towards increasing scientific excellence in the developing world.

But the ICTP shouldn't properly be included here, because it is very likely that the disappearance of UNESCO would not, in fact, make much of a difference to its future. UNESCO currently provides ICTP $507,500 in annual funding through its regular budget: a non-trivial amount for a theoretical physics institute, but hardly a dominant tranche. My guess is that, should UNESCO suddenly cease to exist, the ICTP could probably quite easily find other donors in the public or private sector to fill the gap, given that they actually do important and meaningful work.

Abdus Salam himself was a fascinating figure, whose passing appearance in our analysis very much merits further mention as yet another pointed example of what, fundamentally, is wrong with UNESCO and how, as the saying goes, it never seems to miss an opportunity to miss an opportunity.

A brilliant mathematical intellect from a small market town in the Punjab, Salam not only played a fundamental role in the development of key aspects of the so-called Standard Model of particle physics, he somehow managed to find time for extensive administrative efforts as well, serving as Pakistan's science advisor from 1959-1974, founding its space and nuclear programs, and involving himself with a number of other scientific and administrative initiatives—including, of course, the founding of the ICTP. He was also a devout Muslim and a deeply sensitive and articulate person, arguing ceaselessly for using science as a mechanism for fostering understanding and increasing human welfare through the treatment of problems that were increasingly global in scope.

Like most natural leaders, he was unafraid to tell it like it is, speaking forthrightly and indignantly in his own soft-spoken way.

"There's no such thing as global science as a subject," he once lamented during an interview in the 1980's. *"Even the disappearance of rain forests, which is commanding increased public attention, is not being considered in global terms, as a global asset. People take it as a problem of Brazil, a problem of Malaysia... There is no global vision at all. It's the lack of global vision that worries me, really. It's the issue of globalism which is missing in science, which is missing in the food problem, which is missing in the health problem. What is needed is a vision of a sort I don't see any statesman having."*

But Salam, of course, had such a vision. Moreover, given his strong views and undaunted idealism, it is perhaps unsurprising that in 1986 he tried to throw his hat into the ring to become

UNESCO's Director-General, a hardly unreasonable ambition for an eloquent and passionate Nobel Laureate who had already single-handedly founded a premier UNESCO institute (among others).

But despite his obviously impeccable qualifications and personality, which would have made him one of the few individuals capable of raising UNESCO to a position of immediate global relevance, and despite the active support his nomination would have doubtless garnered from many nations around the globe, Abdus Salam's candidacy was rejected. Indeed, it never even got started, given that Pakistan, his "home country", wasn't the slightest bit interested in nominating him.

Why? Because in 1974, the Pakistani Government passed a law declaring Salam's particular Ahmadiyya brand of Islam invalid and he was henceforth regarded as "a non-Muslim". And in Pakistan, non-Muslims are effectively non-people. And according to UNESCO, if your own Member State won't nominate you, then you are a non-candidate. And so once again, UNESCO structurally ensured itself of resounding non-success. And the band played on.

Education:

UNESCO's Education woes have already been discussed at some length, so there's no point in going into all of that again. It's worth emphasizing, however, that my frustrations here are hardly unique. No less an experienced voice than Nicholas Burnett, an obviously unrepentant idealist who not only served as UNESCO's Assistant Director-General for Education from 2007-2009 but continuously inflicts further punishment on himself by attending numerous "innovative financing" workshops and penning obstinately Utopian sketches of how UNESCO might somehow be reformed[1], has publicly lamented that UNESCO has "...an education staff of very mixed quality", while averring that "it has been slowly deteriorating since the 1970s", losing "intellectual leadership" to both the World Bank and the OECD.

In conclusion, then, none of UNESCO's Education programs stands out as uniquely irreplaceable. Doubtless much admirable work is being done here and there, but that is not, as I have repeatedly emphasized, what we're talking about here. Nothing uniquely exceptional is being done. Nothing. So much for Education.

There is, however, one final cause that should be highlighted, one last argument to potentially resurrect something from the charred UNESCO wreckage: the UNESCO Institute for Statistics (UIS). As an autonomous organization within the UNESCO envelope, its mandate is to provide high level statistical data and analysis for all UNESCO programs; and by all accounts it has managed to establish itself as a serious and competent organization since its founding in 1999. The UIS, however, is clearly a different animal from the other three cases already mentioned (World Heritage Sites, International Hydrological Programme, Intergovernmental Oceanographic Commission) and is largely independent of them. While we've argued that each of the others would likely be reconstituted in some form or another following a sudden UNESCO disappearance, the UIS, being a service facility for a spectrum of UNESCO programs, would only need to be rebuilt to serve the needs of an entirely rebuilt UNESCO. But it's worth pointing out for completeness

---

[1]See, for example, "How to Develop the UNESCO the World Needs: The Challenges of Reform"

that, if UNESCO is ever to be somehow reformed (before or after burning), the UIS would clearly play a substantial role in the construction and delivery of any improved programs.

Chapter 10

8.6% of the Way

This concludes, then, the analytical part of our fiery report: after having seared UNESCO to the ground, we've managed to locate three of its programs—World Heritage Sites (WHS), the International Hydrological Programme (IHP) and the Intergovernmental Oceanographic Commission (IOC)—that would clearly be missed, three distinct programs that we are convinced that the world would feel compelled to reconstruct in some form or another in recognition of their unique importance and impact[2].

In other words, from a fundraising perspective we could perfectly well envision proudly placing all three of these programs in front of any philanthropist, private foundation or government representative while declaring: *"Here are three original, dynamic initiatives that need your support to become even greater still."*

But is that actually the case? Do these three programs really need to be scraping for more resources from third parties? Perhaps the money is already there, squandered

---

[2]In fact, I'm not entirely convinced that the WHS fits into this category and could well imagine a scenario where nobody bothers to reconstruct a neo-WHS from the UNESCO wreckage. But it seemed prudent to be charitable here, given the program's evident popularity.

diffusely through a bucketful of avenues that have long gone up in smoke?

For such are the curious consequences of the fundraiser's theoretical arsonist agenda: all too often, the result of the entire exercise is simply to demonstrate that what is really required is not so much more money in the system, but rather a greater focus, a clearer prioritization of programs. Not always, of course, but sometimes. It's certainly possible. So let's do the math.

In the 2010-2011 biennium budget, the World Heritage Sites Program was allocated a combined $50,350,200 ($15,973,500 in the regular budget, $34,376,700 in extra-budgetary funding). The International Hydrological Programme was awarded $52,867,100 ($29,667,800 in the regular budget, $23,199,300 in extra-budgetary funding) and the Dutch government supplied an additional $73,000,000 for the UNESCO-IHE Institute for Water Education in Delft. The Intergovernmental Oceanographic Commission, meanwhile, received a mere $18,978,800 in biennium support ($10,295,200 through the regular budget, $8,683,600 in extra-budgetary funding).

That is, the combined UNESCO support through the regular budgetary process to all three of these singularly incineration-proof agendas, the shining jewels in the UNESCO crown, is thus $55,936,500, or 8.6% of UNESCO's total annual budget.

8.6%. Think about that for a moment. What could possibly be the reason for UNESCO's flagship programs to receive so little share of the total budget? What is the thinking behind it? What is the rationale?

It could be, of course, that UNESCO's leaders strongly disagree with my reasoning. It could be that their own analysis results in a different priority scheme altogether, a starkly different view of which of their programs are the most unique, the most effective, the most deserving of support.

But no, that's not the case. In fact, that will *never* be the case. Hardly because my reasoning is infallible, but rather because it's painfully evident that this is simply not the way an organization like UNESCO operates: from the Strategic Planning Department to the Executive Board to the Director-General, nobody has the slightest inclination to go through their own brutally honest analysis of which programs are effective and which programs are not, let alone trying to put an enhanced framework into action.

Because for UNESCO officials, the basic problem with UNESCO is simple and it is always about money: there is never enough of it. *Never.* By definition. Whatever the current landscape, through good times and bad, there is always the temptation to harken back fondly to a mythical Golden Age, a time when more resources were available, a time when fewer questions were asked, a time when the world was more open, more sensitive, more progressive. A time when new programs could be started before merrily drifting along unchecked, firmly ensconced in the "UNESCO family" resolutely independent of whatever impact or relevance they might have.

For me, however, my 8.6% analysis had emphatically eliminated once and for all any lingering suspicions that UNESCO's core problem might revolve around a lack of money.

And than it hit me: it wasn't simply that UNESCO's "lack of money" wasn't the problem. It was the solution.

Chapter 11

The Problem

"*The solution?*" I can imagine you raising a sceptical eyebrow here. "*The solution to **what**, exactly?*"

After all, I hardly began my little combustible thought experiment with the idea of realistically developing any sort of solution to what ails UNESCO—my intensely underwhelming experiences from deep in the Strategic Planning heartland having long ago put paid to any delusional fantasies I might have once harboured in that direction.

Indeed, you will recall that when I embarked on this little investigatory voyage, metaphorical match firmly in hand, I quite deliberately treated the whole venture as nothing more than a theoretical diversion to make the best of a bad situation. Instead of promptly scrambling for the exits, I would take advantage of my temporary insider consulting status to become "a modern day Gulliver", travelling sceptically through UNESCO's byzantine bureaucracy to satisfy my curiosity.

And what had I found?

An organization whose quixotically pioneering idealism had inexorably given way to an institutional culture of pathetic mediocrity (at best) fronted by a Director-General with all the intellectual leadership capabilities of a lead paperweight.

A place frozen in intergovernmental stasis, blinded by its own hollow cant and structurally incapable of reform, sliding steadily towards global irrelevance while its panoply of unfocused, largely ineffective programs become increasingly dwarfed by the combined efforts of the World Bank, the OECD, and a whole host of other more effective governmental and non-governmental organizations.

A budget unconscionably tilted towards an obscenely large administrative overhead, where the few remaining substantive or potentially substantive initiatives made up a mere 8.6% of the total operations.

What sort of a solution could there possibly be in these circumstances? Replace my metaphorical arson with the real deal?

In a manner of speaking: yes.

It's time to formally recognize that UNESCO has now reached the stage of being inherently unreformable. Release one's inner Texan. Blow it up. Start over.

But that hardly sounds the slightest bit realistic. Blow it up? How? With what? After all, UN Reform has been talked about for years now and virtually nothing substantial has occurred. And it's not exactly as if, even from a UN-Reform perspective, UNESCO is at the top of everyone's to-do list. Far from it, in fact. If we're going to start blowing things up, why start with UNESCO, of all things?

Ah, but you see, that is precisely my point.

Reforming the United Nations as a whole is admittedly a nearly intractable task. While most rational people have long been convinced of the inherent absurdity of a global governance system whose essential decision-making processes are still fundamentally skewed by the outcome of a war which ended almost 70 years ago, that recognition does very little to change the status quo: inducing any of the five permanent members of the UN Security Council to give up their veto, say, or allow another nation to join their select club in all but the most temporary of fashions.

Effecting meaningful reform at the United Nations is exceptionally difficult to achieve precisely because the UN, for all of its faults, is still widely considered to be important to national interests. Even the United States in its most belligerent and unilateral moments has not seriously contemplated walking away from the United Nations. But it has left UNESCO. Why? Because UNESCO doesn't matter.

By this I don't just mean that UNESCO is perceived as ineffective or inefficient or the agent of initiatives of dubious merit (all of which are clearly true). I mean that the values associated with UNESCO are not viewed as vitally important to American national interests. Or, for the most part, anybody else's.

Now one could lament this. One could decry that the American government, and those of other nations, should be just as worried about global freshwater protection, ocean monitoring[3] or even literacy rates in sub-Saharan Africa than the prospect of imposing economic sanctions on wayward regimes or implementing air strikes. But that is clearly not the case. Which, from our perspective, provides an opportunity.

Because UNESCO's root problem *is* structural: as an intergovernmental institution whose existence is roundly considered a low priority by its more influential Member States, it is in even worse shape than the United Nations proper: suffering the

---

[3]It is worth mentioning that neither the US nor the UK severed their ties with the quasi-autonomous IOC during the years they had withdrawn from UNESCO.

same level of nationalistic politicization that strongly inhibits any meaningful reform without any existing mechanism—such as the Security Council—for the ongoing involvement of the world's most influential countries.

Inevitably, as the major powers increasingly lose interest, UNESCO gradually becomes a focal point only for ambitious, otherwise resentful Member States who cannot hope to make an impact in a wider geopolitical domain, thereby pushing the organization towards an even more politicized orientation on the radical fringe. To this depressing state of affairs, the larger countries respond, inevitably, by sending the lowest members of their diplomatic teams to the increasingly irrelevant UNESCO posts.

To take but one interesting example, Rama Yade, Nicholas Sarkozy's erstwhile Minister of Human Rights who publicly criticized him receiving Muammar Khadafi in 2007 (but didn't resign) was eventually reduced to Minister of Sports for her disloyal outburst. When she again criticized Sarkozy during a controversial speech he gave in Dakar (but didn't resign), she was booted out of the next government and wound up as French Ambassador to UNESCO, where she remained for a grand total of six months before resigning, ostensibly to support the candidacy of Jean-Louis Borloo for the 2012 presidential election (who didn't run). If the French, who at least pay lip service to the importance of the Parisian-based UNESCO, are so manifestly indifferent to its presence, what does that say about its relevance for the rest of the world? The truth is that all the major Western countries have given up on UNESCO a long, long time ago. And everybody knows it.

As night follows day, we eventually reach a point where any further change becomes impossible, as the overly politicized climate in turn ensures an ongoing absence of genuine leadership. Director-Generals become elected through an exercise in diplomatic horse-trading amongst grasping Member States, where the only motivating factor is how best to secure the largest number of significant posts for one's fellow nationals, a process that inevitably favours the selection of diplomats for the top job over accomplished scholars and educators. And should, by some miracle, a dynamic, well-motivated thinker ever somehow find herself in power, it is simply inconceivable that a majority of members on UNESCO's General Council or Executive Board would be the slightest bit motivated to seriously undertake the prioritization process necessary to make even the most basic changes to improve the place.

That is not, of course, the way that things were supposed to have played out.

Back in 1946, Britain and the United States had strongly lobbied for the adoption of a manifestly intergovernmental structure for UNESCO because they were concerned that, without the determined involvement of national governments, it might otherwise evolve towards another version of the anemic, eminently ignorable academic talking shop that its predecessor, the International Committee on Intellectual Cooperation, had become.

France, who had rightly feared the debilitating politicizing effects of such an inherently nationalistic framework on an organization supposedly dedicated to the objective pursuit of education, science and culture had, as we have already mentioned, passionately argued for a non-governmental structure. They lost, but continued to put up a good fight, arguing that, if members of UNESCO's General Conference must be categorized nationalistically, then at least the majority of representatives participating should be chosen from the intellectual community.

This idea was defeated as well, but finally, as a sop to the French, it was agreed that internationally accomplished figures in the arts, sciences and education would be permitted to serve on UNESCO's Executive Board in a personal (i.e. non-nationalistic) capacity. Five years later, even this small attempt at transcending knee-jerk nationalistic bias was formally eliminated as well: henceforth all Executive Board members were to represent national governments. The die was cast[4].

But the French were absolutely right, of course. There is no reason whatsoever to think that the citizens of planet Earth would be well-served by an educational, scientific and cultural organization that is beholden to intergovernmental posturing. Indeed, quite the contrary: anchoring the causes of science, education and culture in a nationalistic context will only reduce their inherent universality and end up completely trivializing their genuine importance.

Which is exactly where we find ourselves today.

Let's cast our mind back to the three uniquely effective UNESCO success stories we discovered through our painstaking process of theoretical flame-throwing: World Heritage Sites, International Hydrological Programme and the Intergovernmental Oceanographic Commission. What do they have in common?

All three are examples of universal programs that successfully develop objective standards to monitor a global good. For the IOC, that good is the state of our oceans. For the IHP, that

[4] It is not a little ironic that today, UNESCO's strongest Western supporter in the teeth of widespread anglospheric condemnation is France, where the organization remains, if not popular, then at least not particularly unpopular. But this probably has much to do with it being located in Paris.

good is the state of the world's freshwater. And for the WHS, that good is the state of our global cultural heritage, as represented by specific geographic locales.

In other words, all three necessarily transcend national proclivities and national politics, which is the principal reason why all three have obtained a measure of success. Few would feel terribly comfortable at leaving the world's freshwater in the hands of the Americans, the oceans in the hands of the Chinese and the list of global cultural landmarks in the hands of the Australians, say. In fact, whatever passport you hold, there is no reasonable scenario where any one country should be in charge of any of these things. In order for these programs to work, an objective, trans-national infrastructure simply has to be created.

But they are, as we have seen in considerable detail by now, hardly typical UNESCO fare. Indeed, they are the exceptions that prove the rule.

Chapter 12

Concrete Utopianism 101

So indulge me for a moment. Let's take our thought experiment just a little bit further.

Let's pretend that UNESCO *does* actually manage to mysteriously burn down to the ground and the UN Secretary-General, unaware of my suspicious, albeit theoretical, arsonist tendencies in matters UNESCAN, but hearing that I have thought deeply and seriously about structural matters in my capacity as a short-lived strategic fundraising consultant to the once-mighty Bureau of Strategic Planning and anxious to seize this opportunity to boost the reputation of what had long been regarded as the most hidebound and ineffective of all UN agencies, spontaneously invites me to head up a one-man exploratory committee known as BURP (Bold UNESCO Replacement Program) to consider how best to concretely move forward with UNESCC II.

"*Think big,*" the Secretary-General urges me with an encouraging smile. "*I'm open to anything. Just tell me how to make it better.*"

So, after careful analysis, what, specifically would I recommend?

Well, I'd begin by simply defining the primary focus of the place: to develop innovative, unique and highly effective programs in education and science.

Gone would be the florid phraseology about constructing the defenses of peace in the minds of men (after an exhaustive 65-year study, we can safely conclude that UNESCO's overall impact at promoting peace in the minds—or anywhere else—of men—or anyone else—was effectively zero).

Gone too would be all mention of humanism (Renaissance, Byzantine, Christian, secular, civic, evolutionary, global, total or otherwise), "bright zones", "dark zones", eugenics and all the rest, along with the problematically vague and politically charged word "culture". Just science and education: a big enough, and important enough, terrain to get one's teeth into, one might think.

And then I'd need a name for the thing. How about the United Nations Innovative Science and Education EXplorations? In other words, UNISEX—a reflection of the gender-neutral, universal importance of both science and education to our world. One size fits all.

UNISEX would strive to quickly establish itself as a science and education programmatic innovator and unique content provider, simultaneously collecting data on a whole host of scientific and educational matters in our capacity as a rigorously objective UN Agency, while sharing our specifically-designed programs with any region, nation or national collective that was motivated to work with us. All national governments and regions would be welcome, but none would be in any way obliged.

What sorts of things might UNISEX do? Here are a few obvious ideas to get started:

Provide a wealth of pedagogic materials to educators the world over to tangibly assist them in the classroom. Generate extensive interactive support networks for teachers to exchange best practices, techniques and emotional support. Work with existing schools, universities, foundations, NGOs throughout the world on exchange programs and shared resources.

Construct a wealth of informative and inspiring content (interviews with celebrated figures, educational applets, eBooks, etc.) to motivate students and teachers alike. Create social networks to enable motivated youth from the developed world to regularly participate in UNESCO educational projects, both online and in person. Develop regional programs where local success stories are specifically presented as inspiration to the next generation and have the opportunity to regularly interact with vast numbers of young, impressionable students.

Produce a spectrum of educational games and multi-media resources to clearly articulate the importance of biodiversity, complete with concrete examples of systemic failures. Fully harness the analytical resources of the Institute for Statistics to develop and implement a wide spectrum of PISA-like testing for all countries of the world that chose to participate, specifically measuring enrolment rates (for both boys and girls), academic achievement, adult literacy, and a host of other objectives. Use the website to proudly and explicitly trumpet the successes of those countries or regions that have reaped the fruits of the available programs. Partner with other NGOs and foundations to recycle hardware and software resources from the rich world to schools in impoverished areas.

The list is virtually endless. But the important thing is to focus on developing specific content combined with UNISEX's unique modes of delivery. And any program that is developed must be consistently reviewed to ensure that it is still effective, modern, relevant and done to the highest possible standards.

Then I'd talk about structure.

UNISEX would be a special type of new United Nations hybrid: an autonomous, international body formally under the UN umbrella, but small and flexible and strongly dependent on non-governmental experts like the original French recommendations of so long ago, convinced as we are from the UNESCO experience that the ideals of an effective science and education body are best served by an organization independent of the political motivations and machinations of Member States.

It would have a small governing board (say nine members) incorporating impressively accomplished members of the international philanthropic, scientific and educational establishment. People like famed Somali gynecologist, educator and human rights worker Hawa Abdi, Indian philanthropist Anu Aga, French doctor Philippe Douste-Blazy, American-Canadian physicist Nima Arkani-Hamed, Nobel Prize-winning Economist Elinor Ostrom and American mathematician and financier James Simons. Add Google's Sergey Brin and Columbia University's Jeffrey Sachs, perhaps. There is no shortage of suitable candidates. The UN Secretary-General would sit on the board ex-officio and chair the board, but only vote on resolutions in the event of a tie.

UNISEX would be committed to achieving a broad diversity of governing board members in terms of gender, geography and professional experience, but would never formally invoke a quota system which could result in a temporary sacrifice of intellectual quality of potential candidates. The normal board tenure would be three years (non-renewable), but owing to the logistics of setting up the structure, the original nine board members would have randomly selected staggered terms of 2, 3 or 4 years, so that a steady-state situation is reached after only two years: thereafter, three board members would be replaced each year.

By establishing a serious international organization of the highest order, there would be no shortage of top-level thinkers who would be willing to volunteer their time to ensure UNISEX's success. And the more successful and prestigious UNISEX becomes, the easier it will be to recruit top-level people to its governing board.

But the governing board is not actively in charge of UNISEX. It serves three primary functions:

1. Oversight
2. Selection of a suitable Director-General
3. Assistance to the Director-General

Oversight involves not only ensuring the absence of any improprieties, but also that UNISEX maintains an appropriate level of transparency for the outside world. To be maximally successful UNISEX not only has to be an open institution, it has to be seen to be as such.

The recruitment of a top quality Director-General is UNISEX's top priority and thus represents the single most important task facing the founding board of directors. It is essential to attract a strong, dynamic, articulate Director-General to lead the organization and be personally held accountable for its successes or failures.

Where to start? Extreme professional accomplishment is necessary, but hardly sufficient. Just possessing a Nobel Prize, for example, however impressive that might be, is hardly a guarantee of success for the varied, subtle, task awaiting the new Director-General of UNISEX. She must possess unimpeachable credentials and considerable perspicacity. It is vital to be both analytical and strategic, passionate yet sensitive, equally adept in a board meeting as in a laboratory. Such people are not plentiful. But they certainly exist. Abdus Salam, as we've already mentioned, would have been just what the doctor ordered.

But there are certainly others alive today: someone like entrepreneurial biologist Craig Venter, worldwide web inventor Tim Berners-Lee, Cambridge physicist Athene Donald or Nobel Laureate Elizabeth Blackburn. Notwithstanding the challenging combination of desirable characteristics, the prospective field is still quite sizable, but only if one proactively searches rather than waiting for applicants to apply themselves. All too often it's the very people who *aren't* angling for these sorts of jobs in the first place—who are presently busy with many of their own transformative ideas and initiatives—who would make the best candidates, people who are much more concerned about changing the world rather than simply buttressing their CV with another title.

Then there's the question of where to put UNISEX headquarters. It's high time to move it out of Paris, which is only a distraction (albeit it a highly enjoyable one). Given its mandate, it makes far more sense to place the head offices in one of the BRICS (Brazil, Russia, India, China or South Africa), with my personal preference being South Africa as public recognition of having the conspicuous probity to avoid a bloody civil war post-apartheid.

It would have a very limited number of regional offices (perhaps four or five: Brazil, Mexico, India, China, Russia) that would ideally be shared with other UN agencies, and the Director-General and her senior management team would be strongly encouraged to make regular trips to interact with interested and motivated partners around the world, leveraging existing UN infrastructure.

There would be annual, public reviews of the institute's performance. Only one prize would be given annually, announced immediately after the Nobel Peace Prize with considerable media attention, *The UNISEX Award for Innovative Education*, with a monetary value of exactly the same as the Nobel Prize.

Lastly, UNISEX would regularly strive to ensure that it was working as closely as possible with other government agencies, philanthropic foundations, NGOs and so forth to avoid redundancies and maximize its global impact. It would promote a culture of unflinching transparency and the highest ethical standards, recognizing its elevated status as a global scientific and educational role model. As mentioned previously, UNISEX would be consistently determined to maximize the new information and communication technologies to develop and maintain an ongoing dialogue with all peoples of the world, developing its website into a global portal for captivating scientific and educational content for teachers, students, researchers and the general public.

And what about our erstwhile UNESCO stars that we discovered in our previous analysis? What would happen to them?

We've already seen how UNESCO's Institute for Statistics (UIS) would be directly incorporated within UNISEX. Why not have the International Hydrological Programme (IHP) follow suit, with a spectrum of concrete educational materials and programs developed that specifically resonate with its unique mandate and accomplishments.

The Intergovernmental Oceanographic Commission (IOC), on the other hand, should finally be established as a separate UN Agency with its own founding charter, working closely (as it naturally does already) with the United Nations Environment Program (UNEP), World Meteorological Organization (WMO) and other relevant bodies. While the Member State structure is far from ideal, as we've repeatedly seen, it does make a good deal of sense in some specific, limited contexts.

For the IOC, which is so involved in a number of specific, meritorious capacity-building objectives (coastline protection, environmental sensitivity, tsunami early warning systems) as well as broader-based ocean management and governance, there is no a priori reason why its interests couldn't best be served by establishing itself as a stand-alone UN agency under the standard Member State model, and considerable reason to fear that, owing to the breadth of issues under its charge, it might naturally dilute UNISEX's focus.

That being said, just like IHP, UNISEX should make a special effort to work directly with the IOC to design and implement a vast array of related pedagogical programs and materials to leverage each others' strength to their mutual benefit.

Then there's the World Heritage Sites (WHS) program.

"Why not seize the opportunity to do something truly innovative here?" I can imagine myself smilingly suggesting to the Director-General.

Why not initiate a brand new United Nations program in participatory global democracy? The idea would be to announce that some finite limit—say, one hundred—of the most beautiful and culturally significant locales would be annually selected through an internet ballot by all the peoples of the globe.

Of course, there would be some difficulties at first—an obvious point of concern is that it's hardly the case that everyone on Earth has access to the internet. On the other hand, however distorting that state of affairs is, it's not clear that the status quo—where some guys sitting around a table in a Paris building decide who gets to be on the World Heritage Sites list—is so much better. At least doing things this way is a concrete step in a more democratic direction and could well serve as some (albeit minor) further incentive to increase global access to modern communication technologies.

Not only that, but the cost of the whole program would drop drastically from its current levels. Recall that at present the WHS program requires an annual allocation of slightly more than $25 million to run (about $8 million through the regular budget and $17 million in extra-budgetary funding). By switching to an internet ballot, we would likely only need a few website guys and a couple of software engineers that we could probably set up in some of the off-site offices of former UNESCO delegations that managed to somehow escape our UNESCO inferno unscathed.

Of course it would take some time to get the bugs out of the software. In particular, there's the thorny issue of ensuring that determined internet voters don't somehow discover a way to vote more than once for their favourite World Heritage Sites. But unlike similar low-tech ballot stuffing that occurs throughout many parts of the world, at least here we have the peace of mind of knowing that such electoral transgressions don't have any real effect whatsoever.

Chapter 13

Harnessing A Crisis

Thanks for your indulgence.

But the scepticism, of course, remains. Stronger than ever, in fact.

"Yes," I imagine you admitting, "I must confess that I've hardly been preoccupied with UNESCO in the past, but after listening to your arguments, I'm now quite convinced that we'd all be much better off if UNESCO could somehow be magically transformed into something quite different: focused, efficient, apolitical, analytical, impactful. Yes, it should have a lean and mean management structure, with an executive board of dedicated and extremely experienced individuals anxious to see objective results and determined to hold the leadership accountable. Yes, it should have senior leadership at the highest possible international level. Yes, it should harness the internet and new media to the fullest possible capacity as a means of disseminating a wealth of unique high level content to educators, policy makers, students, researchers and the general public. And yes, we should probably take it out of Paris and put it somewhere new—quite possibly in a place like India or Brazil or South Africa.

"Yes, yes, yes. *But so what?*

"After all, it is one thing to diagnose the problem and even come up with a theoretically viable solution, but it is quite another to actually be able to implement it. The first two are, in fact, the easy bits. Knowing what has to be done is hardly the same as actually being able to do it."

"Ah," I would reply, "but it turns out that I *do* know how to do it. I *do* know how to bring about the change—or rather implosion—that we need. You see, it's really quite simple: it's all about the money: just take it away."

In other words, it's time to practice some truly innovative fundraising. I call it: *fundlowering*.

Organizations like UNESCO typically need a worldwide crisis to bring them in and out of existence. It took WWI to create the League of Nations and the International Committee on Intellectual Cooperation. It took WWII to bring them out of existence and develop their updated versions, the United Nations and UNESCO. Obviously the goal here is not to create WWIII. However much we are despondent at what UNESCO has become, it's awfully hard to justify that.

But wars do not represent the only sort of worldwide crisis we can be faced with. When it comes to crises, one can imagine a depressingly wide variety of possibilities: moral crises, spiritual crises, political crises. Most significantly of all, every so often a global economic crisis rears its ugly head, leaving a trail of unemployment, budget cuts, dried up credit and worldwide anxiety in its wake.

And it so happens that we are in the middle of one. Let's constructively use it to our advantage by convincing Member State governments to stop wasting their citizens' hard-earned resources on UNESCO's biannual budgetary contributions and spend it directly on themselves instead. It's not as if there aren't any important domestic causes that could benefit from some added resources. How about the homeless? Or improved health care? Or government retraining programs for the unemployed? It's not a huge amount, of course, this UNESCO contribution, but it's not insignificant either. Every little bit helps.

Doubtless UNESCO supporters would scream and cry that withdrawing Member States' contributions would be a slap in the face to the principle of multinational cooperation, but this is, of course, quite incorrect. Since the goal is to capitalize on a financial crisis to bring about a substantially improved and effective UN organization, withholding UNESCO contributions is only a slap in the face to the principle of *pointless* multinational cooperation, or perhaps *counterproductive* multinational cooperation, which are very different principles indeed.

It's important to bear in mind that money is UNESCO's lone Achilles' heel. It has no real image to protect, no shining reputation that might be somehow sullied. It is impervious to wry, sarcastic commentary on its profound irrelevance from the academic, corporate, philanthropic and even broader intergovernmental communities. That has been going on for years now, to no apparent effect. Even most Member States know this: there is a reason, after all, that UNESCO budgets have been frozen for years, not even keeping pace with inflation. But things just drift on and on. Everyone is afraid to pull the plug.

They shouldn't be. If UNESCO were to be seriously hit in the pocket book it would eventually totter and collapse under its own weight, falling like a booby-trapped AT-ST Walker during the Battle of Endor, while we, joyous citizens of planet Earth, would be dancing and celebrating like the victorious Ewoks vanquishing the perfidious Empire forces, delighted to finally have the opportunity to create a meaningful United Nations body for science and education that the world so desperately needs.

"But," you might well ask, "Star Wars imagery aside, let's be realistic for a moment. Once you've forced UNESCO to disappear by withholding Member State funds, where will you find the money for UNISEX?"

This is, of course, an important question—but once again addressing it *now* is simply the wrong way to go about doing things, irrevocably leading us down the unproductive path where content and impact take a disturbingly distant second (or worse) place behind the great obsession of constantly scurrying after any available financial resources.

This we will not do. For UNISEX has learned from the UNESCO experiment where balancing the books had become completely decoupled from the merits of its panoply of unfocused programs. Never again will our fundraising efforts be done in a void, never again will we demand budgetary satisfaction as some kind of international moral birthright. No: UNISEX will be all about *content*: what it can do, what it should do, what it must do to deliver on its focused mandate of developing innovative, unique and highly effective programs in education and science. Of course this will require funding. And sometimes, we will naturally be faced with a situation where we need to find additional sources to extend worthy programs or keep existing ones alive.

But the content comes first. **First** we think about what needs to be done, **then** we itemize about how much all of this will cost, **then** we worry about where the money comes from. That's the UNISEX way.

And if we stick to our guns here, we'll soon discover that there are, in fact, lots of reasons for confidence on the financial front. After all, it's not like the international community is unwilling to participate in our cause. As mentioned previously, UNESCO's annual regular budget from Member State allocations is $326,500,000. That could create a whopping amount of impact, if used wisely.

And if Member States are willing to spend this much for a litany of almost wholly ineffective programs, it stands to reason that it wouldn't take much to convince them to spend at least the same amount for something actually meaningful. To first approximation, one could easily imagine many Members States simply bundling the amount they formerly paid to UNESCO as part of their general UN allocations and having the UN accountants simply transfer the total to UNISEX directly. Yet another reason to have the UN Secretary-General serve as chair of the board of UNISEX.

And then there's the whole vitally important question of how much UNISEX would actually need in the first place. It's hard to believe it would actually require the full annual budget of $326,500,000 that UNESCO currently receives for it to do things properly, at least not at the beginning. There's no way it should require more than half of its funds for administration, of course. More specifically, it certainly wouldn't need UNESCO's current annual tranche of $22,000,000 for "General Policy and Direction" (including $2,740,600 for the General Conference and a whopping $10,178,050 for "Direction"). The current UNESCO line item of $5,482,600 for "Participation in the joint machinery of the UN system" could obviously be avoided—indeed, it's hard to see what on earth this is all about.

The office of UNESCO's Director-General currently has an annual budget of $3,205,250, and one wonders what this is all for. Clearly not much goes to the creation of grand ideas. Or hiring leaders with any relevant domain expertise. Or even fact checking for speeches.

As we've mentioned, the Director-General is a pivotal position for UNISEX and all efforts must be made to ensure that UNISEX's DG has sufficient resources to put the new institution firmly on the international map. But surely this could be done with only about half of what UNESCO currently spends on the Director-General's Office while simultaneously gaining a much bigger bang for the buck. After all, what more does a Director-General really need than a good secretary and a hefty travel budget? If you make sure to get a really good one and be prepared to pay top dollar for her (let's say $1 million, just to take a nice, round, borderline outrageous figure), you can also assume that she will write her own speeches. And actually know what she's talking about.

And while we're on the subject of making salaries sufficiently attractive, it's worth emphasizing that Member States also directly contribute to UNESCO by allowing all salaries of UNESCO full-time staff to be tax-free, thereby, at least in theory, greatly increasing the organization's potential competitiveness to recruit top people.

Presumably this dispensation would be awarded to UNISEX staff as well.

Other obvious mechanisms suggest themselves. Rather than make a distinction between budgetary and extra-budgetary contribution as exists today, we can simply allow for a (fully transparent) opportunity for any willing country to directly enhance its participation in a particular UNISEX program by directly scaling up their own contributions. So if country X finds that a certain UNISEX program is proving to be particularly effective and would naturally like to see it extended towards a wider region than originally intended, it will be able to clearly add to that extension in a clear, transparent and open way that will also serve as a public endorsement of UNISEX's established initiative and will hopefully result in other countries and regions following suit, thereby enabling the program to have a still greater impact.

As previously mentioned, UNESCO's current "extra-budgetary" process tends to work in the other direction: towards a dilution of original programmatic focus by introducing one-off initiatives that naturally tend to be more aligned with the interests of the ruling local governments than UNESCO's basic programs. Unlike UNESCO, then, UNISEX won't make a habit of being "open for business" by wantonly rushing after whatever resources any particular government is motivated to throw at it for its own often nebulous purposes, tailoring programs and policies on the fly and becoming increasingly incoherent.

Programs will be reviewed in the normal course; and during that review process all input from all individuals and government representatives—whether active participants or not—will be welcome. But once a program is established it will run its course, and local and regional governments will be left with the sole option of either involving themselves in it or not.

In short: programs won't follow the money; money will follow the programs. Aside from streamlining administrative expenses, this approach will greatly aid UNISEX in establishing both its international credibility and its associated brand. At the end of the day, if done right, those will translate into as much money as will ever be needed.

But active participation with the public sector won't represent the only strong relationships that UNISEX will have. There is also the increasingly relevant world of NGOs, foundations and private philanthropy, not to mention developing active partnerships with additional like-minded institutions, such as universities. A dynamic and strategically-oriented UNISEX will be able to constructively avail itself of many additional resources from this quarter as well, given the universal concerns of avoiding duplication, sharing resources and integrating programs to have a maximum joint effect.

So more money or in-kind support will come from this direction as well. But developing meaningful relationships with other similar organizations is not just about financial resources. Successful philanthropists often have different perspectives and hard-earned expertise garnered from a lifetime in the private sector that UNISEX would be well-advised to benefit from, and there is a lot to learn from a whole range of effective private foundations and NGOs that have been around the block a few times.

That is why I would strongly recommend that several high-profile representatives from the international philanthropic and NGO community sit on UNISEX's governing board (this would also, I imagine, have the added bonus of eliminating, or at least seriously diminishing, the annual $3,912,100 that UNESCO currently spends on its Executive Board. Where precisely does this go, I wonder?).

Moreover, for obvious structural reasons there are often times when a non-governmental organization or private foundation can more successfully interact with local government representatives than a UN body, however apolitical and flexible its orientation might be. An astute UNISEX might often be able to use international partners as independent "shields" in sensitive geopolitical areas of the moment so as to most constructively maintain and extend its programs.

In short there's no need to worry about where UNISEX might find its required financial resources to do its job properly, assuming it actually does so. The problem is not about money. In fact, the problem has *never* been about money. There are all sorts of ways of raising significant sums of money for the right causes and UNISEX's job is to focus on ensuring that it has the best programs and the best structure and there is every reason to confidently believe that the money will clearly follow.

But first, of course, UNESCO—its counterproductive evil twin—must be destroyed. And that, as I've already mentioned, *is* indeed all about money. For the only way to finally drive a stake through the UNESCO heart once and for all is through its pocketbook.

The good news is that it is already happening. But, in typical UNESCO fashion, it is happening for entirely the wrong reasons.

On October 31, 2011, directly following a UNESCO vote to formally admit Palestine to the UNESCO fold, the United States announced that it was going to withdraw its financial support, which amounts to some 22% of the regular UNESCO budget.

This announcement was followed, unsurprisingly, by Israel's decision to suspend its financial contribution (roughly $2 million) and Canada's rather more tepid response of freezing all voluntary (i.e. extra-budgetary) contributions (this is the ***country*** of Canada, you understand, as opposed to its ***province*** of Québec, that in 2006 was granted independent UNESCO membership as a sop to Québecois separatists, further proving that once one begins to tread down the turbid road of melding political appeasement with international development there is no telling which irrelevant parochial squabble might eventually be brought into play).

That the United Nations Education, Science and Cultural Organization has deliberately elected to engage itself as the scene of a proxy war for the Arab-Israeli conflict is a perfect example of just why, precisely, it must be immediately shut down.

Not because the Palestinians don't deserve a state of their own. Or because the Israelis don't deserve to have sufficient security guarantees. Or as a commentary on current Israeli settlement policies in the West Bank. Or as a referendum on the Palestinian leadership. Or as a reaction to the strength of "the Jewish Lobby" in Washington. Or as part of the political calculus of Barack Obama's re-election. Or to appease the Republican majority in the House of Representatives. Or the prospect of Iran building a nuclear bomb. Or any of that.

No, the obvious point to be made is simply this: none of this should properly have the ***slightest thing*** to do with a UN education, science and culture agency *whatsoever*.

The fact that it clearly does, the whole idea that UNESCO's prospects for maintaining its incoherent spectrum of programs and its bloated administrative overhead is now inextricably related to the intractably muddled geopolitics of the Middle East, is the most conclusive evidence possible that something has gone terribly, irretrievably wrong. You don't even need my detailed 8.6% argument at all any more. Just shut it all down.

Right now.

Not because you support Israel. Not because you support Palestine. But because you support the inherent values of science and education that should properly lie at the heart of this institution. And because it has become crystal clear that, for any international organization centred around the values of science and education, the Member States approach not only doesn't work: it is nothing less than a recipe for disaster.

In other words, the experiment has clearly failed. The French, it seems, were right after all (they were due, after all—they've had to suffer the likes of Henri Bergson). By forcing the United Nations education and science agency to structure itself directly on the interests of 196 Member States, we have finally reached the point of no return: a bureaucratic miasma of frightening proportions where programmatic ad-hockery reigns supreme, Human Resources is run on strictly nationalistic grounds, leadership is non-existent and the entire establishment is held hostage to periodic geopolitical flare-ups that are wholly irrelevant to what its core mandate should be.

Enough. Rather more than enough, in fact.

It's time to demonstrate to the world that, as befitting an institution devoted to the causes of science and education, we can objectively assess our own past and learn from our mistakes in order to improve. It's time to be scientific, in other words, about our own education. It's time to practice what we preach.

So there we have it: my final conclusion, the end product of months of diligent research:

The proper way to ensure that there exists a vibrant and effective United Nations agency that can readily deliver an ambitious science and education agenda to the international community is to do one's utmost to ensure that, in the short term, the present UN agency affiliated with this mandate—UNESCO—receives *no more money whatsoever*. Not one penny. Pas un centime.

In other words, what UNESCO desperately needs is not fundraising. What is now required is *fundlowering.*

Doubtless this is not what Hans D'Orville had in mind when he hired me. But you have to admit that it ***is*** innovative.

And it just might work.

Epilogue

Hope Springs Eternal

The US and Israel formally left UNESCO in December 2018, after having announced their intention to leave the previous year, citing anti-Israel bias.

While some will doubtless unhesitatingly ascribe such a decision solely to the ultra-isolationist tendencies of the Trump administration, it's clear that doing so would be a mistake. As mentioned in the previous chapter, UNESCO's 2011 decision to admit Palestine as a fully-fledged member of the organization promptly led to a freezing of American funding by the Obama administration, and the writing was clearly on the wall for a subsequent American withdrawal, which UNESCOvites seemed perversely determined to trigger, continually wading into the Middle Eastern political fray and promulgating various Palestinian sites and attractions with an unmistakable whiff of anti-Israel bias. It is hard to believe that a Democratic administration would have continued to maintain UNESCO membership under the circumstances, just as it is hard to imagine Israel would remain after the US withdrew for perceived anti-Israel bias.

All of which simply re-expresses the very depressing status quo that an institute whose sole purpose for existing is to increase the global awareness and appreciation of science, education and culture is justifiably regarded as nothing more than a third-rate vehicle for spewing intolerance and sociopolitical aggrievement.

Irina Bokova spent eight years as UNESCO's Director-General before moving onto other activities, such as a teaching post at Korea's Kyung Hee University, where, one presumes, she diligently engaged in educating the next generation of Koreans about how Denis Diderot, Hannah Arendt and Erwin Schrödinger collectively transformed the intellectual and cultural landscape of late 17th-century Albania.

The three organizations specifically recognized as meritorious UNESCO-affiliated initiatives—World Heritage Sites (WHS), the International Hydrological Programme (IHP) and the Intergovernmental Oceanographic Commission (IOC)—still remain within the UNESCO fold, with the WHS, as always, being the only one that anyone outside of the world of international development has ever heard of, despite the dramatically increasing global need for ocean stewardship and international water research and management.

So far, so depressingly predictable.

But there is, rather remarkably, a spark of unexpected good news—great news, even, as these things go. UNESCO's current Director-General is the eloquent and breathtakingly competent Audrey Azoulay, who has not only graduated from various well-recognized institutions (including the very prestigious École nationale d'administration) but also briefly served as the French Minister of Culture.

I wonder if she has an extra lighter lying around somewhere. More interestingly still, I wonder if she has the temerity to use it.

www.ingramcontent.com/pod-product-compliance
Lightning Source LLC
Chambersburg PA
CBHW030908080526
44589CB00010B/210